Simulating Clastic Sedimentation

COMPUTER METHODS IN THE GEOSCIENCES

Daniel F. Merriam, Series Editor

Computer Applications in Petroleum Geology
Joseph E. Robinson
Graphic Display of Two- and Three-Dimensional Markov Computer Models in Geology
Cunshan Lin and John W. Harbaugh
Image Processing of Geological Data
Andrea G. Fabbri
Contouring Geologic Surfaces with the Computer
Thomas A. Jones, David E. Hamilton, and Carlton R. Johnson
Exploration-Geochemical Data Analysis with the IBM PC
George S. Koch, Jr. (with programs on diskettes)
Geostatistics and Petroleum Geology
M. E. Hohn
Simulating Clastic Sedimentation
Daniel M. Tetzlaff and John W. Harbaugh

SIMULATING CLASTIC SEDIMENTATION

DANIEL M. TETZLAFF, Western Atlas International
JOHN W. HARBAUGH, Stanford University

VNR VAN NOSTRAND REINHOLD
New York

Publisher's note: This is the last book in the series to be published by Van Nostrand Reinhold.

Softcover reprint of the hardcover 1st edition 1989

Library of Congress Catalog Card Number: 88-12003
ISBN 978-1-4757-0694-9 ISBN 978-1-4757-0692-5 (eBook)
DOI 10.1007/978-1-4757-0692-5

Van Nostrand Reinhold
115 Fifth Avenue
New York, New York 10003

Van Nostrand Reinhold (International) Limited
11 New Fetter Lane
London EC4P 4EE, England

Van Nostrand Reinhold
480 La Trobe Street
Melbourne, Victoria 3000, Australia

Macmillan of Canada
Division of Canada Publishing Corporation
164 Commander Boulevard
Agincourt, Ontario M1S 3C7, Canada

16 15 14 13 12 11 10 9 8 7 6 5 4 3 2 1

Library of Congress Cataloging-in-Publication Data
Tetzlaff, Daniel M., 1954–
Simulating clastic sedimentation / Daniel M. Tetzlaff, John W. Harbaugh.
p. cm. — (Computer methods in the geosciences)
Bibliography: p.
Includes index.
ISBN 978-1-4757-0694-9
1. Sedimentation and deposition—Computer simulation.
I. Harbaugh, John Warvelle, 1926– . II. Title. III. Series.
QE571.T44 1989
551.3—dc19 88-12003
CIP

Contents

Series Editor's Foreword

Simulation. "Why simulate?" we might ask. Dan Tetzlaff and John Harbaugh answer that question in *Simulating Clastic Sedimentation,* and answer it effectively. Readers will have no trouble following this well-documented, clear exposé on clastic-sedimentation simulation. The sophistication in simulation since 1970, compared to the early work, is enormous. The development and availability of hardware, especially graphics, has made such advancement possible. The authors' presentation allows us to share in this exciting new approach to geological problem solving. The book contains a glossary, notation convention, program listings, plenty of illustrations, and chapter summaries to enlighten the users. Everything is readily available to make this a self-contained, how-to and how-to-interpret book.

After introducing the subject of simulation, the authors take the reader step-by-step through the computation and reasoning of the erosion, transportation, and deposition of clastic sediments. Fluid flow, transport, boundary conditions, and space and time are all explained, and details are given as to how each of these factors were accounted for in the simulation. The simulation is compared to different known situations, such as braided streams and deltas, for assessment as to realism. The final chapter gives two real-world applications where a simulated model is compared to the Simpson Canyon Field in northern Alaska and the Golden Meadow Field in southern Louisiana. From these simulations was gained some understanding of the conditions under which certain sedimentary units were formed. From this understanding, then, better predictions could be made of the present distribution and preservation of units. And, after all, accurate prediction is the ultimate goal.

The authors are well-qualified to bring the material to us. Tetzlaff, a former student of Harbaugh, developed the technical aspects of the study and implemented them. Harbaugh himself has been involved with simulation studies for 25 years and, with Graeme Bonham-Carter, wrote the definitive text on the subject (Wiley, 1970).

This is the seventh book in the series on Computer Methods in the Geosciences since its inception in 1982. Topics have included such diverse subjects as Markov models, image processing, spatial analysis, simulation, and geostatistics as applied to petroleum geology, remote sensing, and geochemistry for use with microcomputers to superminis. Depending on the user's background, *Simulating Clastic Sedimentation* can be used as a text or a reference. Whatever the level, it is a "must" for the forward-looking student, practitioner, or researcher.

D. F. MERRIAM

Preface

The main thrust of this book is to provide a new way of interpreting the history of the earth's sedimentary basins. We take the view that the geologic processes that have shaped the Earth's crust can be represented by simulation models, and that these models can be used to mimic the development of the Earth's crust through geologic time. This book is about those specific processes that erode, transport, and deposit grains of sediment consisting of sand, silt, and clay. Common as these processes are, they have been of immense importance in transforming the Earth's crust over the past 4 or 5 billion years.

The term *simulation model* has various meanings. The models described here represent geologic processes acting in three dimensions through geologic time. In this context, simulation provides a means of experimentation in which specific geological experiments are carried out by establishing initial conditions and then setting the model to work. The model moves forward through time, rivers and marine currents flow, sediment grains are transported and deposited, and sedimentary basins evolve. Given sufficient time and geographic expanse, the depositional history of an entire sedimentary basin can be represented in this fashion.

The simulation models presented here consist of computer programs that direct a computer's operation. The flow of water and the transport of sediment are represented by equations and logic operations in the programs. When a major experiment is carried out, as many as a trillion arithmetic and logic operations may be required.

This book expresses our conviction that the evolution of the Earth's features can be treated in terms of the processes that created them. Although this point of view has been widely adopted by geologists, we have extended it by devising procedures that represent processes rigorously and quantitatively, and permit them to be operated consistently through thousands of years of simulated geologic time.

The book provides a description of equations that represent the flow of water and its capability of eroding, transporting, and depositing grains of sediment. We also describe how these equations are represented in a form that permits them to be solved numerically by computer programs. We have stressed graphic display, emphasizing that results of simulation experiments must be graphic if they are to be effective, and our description includes details of the computer display procedures.

We hope that this book will appeal to geologists with widely ranging interests. Although we have focused on the specific processes of flow and sediment transport, the book's general philosophy and the procedural methodology are broadly applicable. Our overall intent is to stimulate the general application of process simulation in geology.

ACKNOWLEDGMENTS

We are grateful to Texaco, Inc., for providing the financial support for much of this study. Donald F. Beaumont, formerly of Texaco, and Richard E. Byrd, Michael Zeitlin, and Martin Perlmutter of Texaco were supportive in obtaining funding as well as in providing ideas and encouragement. We acknowledge the support of Charles Hutchinson, formerly of Van Nostrand Reinhold, for early encouragement in producing this monograph.

In 1986, Gould, Inc., provided a PowerNode 9080 computer system, on which most of the computer-intensive runs were made. The high speed and versatility of the 9080 computer allowed us to perform simulation experiments that would have been impractical on less powerful computers. We are indebted to James Enking, Stephen Coles, Pat Rickard, and Robert Leaf, among others at Gould, who helped arrange the donation and the maintenance program for the PowerNode 9080. Earlier we used a Gould/SEL 32/77 computer donated by ARCO Exploration Company to Stanford University and subsequently maintained by Gould.

We are indebted to our colleagues at Stanford, who include Professors David Larue (now at the University of Puerto Rico), Stephan A. Graham, Joseph Franzini, and Jeffrey Koseff. Dr. Ralph Cheng of the U.S. Geological Survey helped with suggestions in early stages of the study.

We thank Stanford students Robert Laudati, Paul Martinez, Norman Scott III, Luis Ramos, Phillip Kushner, Claudio Bettini, and visiting scholar Klaus Bitzer. We acknowledge Alicia I. Tetzlaff, who assisted in typing part of the original manuscript, and Mario Rossi, who helped us so extensively in numerous subsequent revisions.

DANIEL M. TETZLAFF
JOHN W. HARBAUGH

Simulating Clastic Sedimentation

CHAPTER 1

Introduction

The initial idea for this study began in the 1960s, when simple, two-dimensional dynamic computer simulation models of shallow-water marine sedimentation processes were developed at Stanford. Some three-dimensional simulation models were considered then, but they posed demands that were too great for the computers then generally available, or at least too great for those that we could afford. The work languished for about 15 years, but in 1982 it was resumed when we began work on a coordinated series of quasi-three-dimensional computer simulation modules (Fig. 1.1) dealing with geologic processes that create sedimentary basins. The resumption was stimulated in part because we could then afford computers large enough and fast enough to perform the computational labor that three-dimensional process simulation requires.

The first module in this series is named SEDSIM, and it simulates erosion, transportation, and deposition of clastic sediment. It can be used in concert with the other modules shown in Figure 1.1 that are currently being developed. There are three versions of SEDSIM:

1. SEDSIM1 deals with fluid flow only.
2. SEDSIM2 deals with flow and with transport of clastic sediment of a single particle size.
3. SEDSIM3 deals with flow and with transport of clastic sediment of up to four particle sizes.

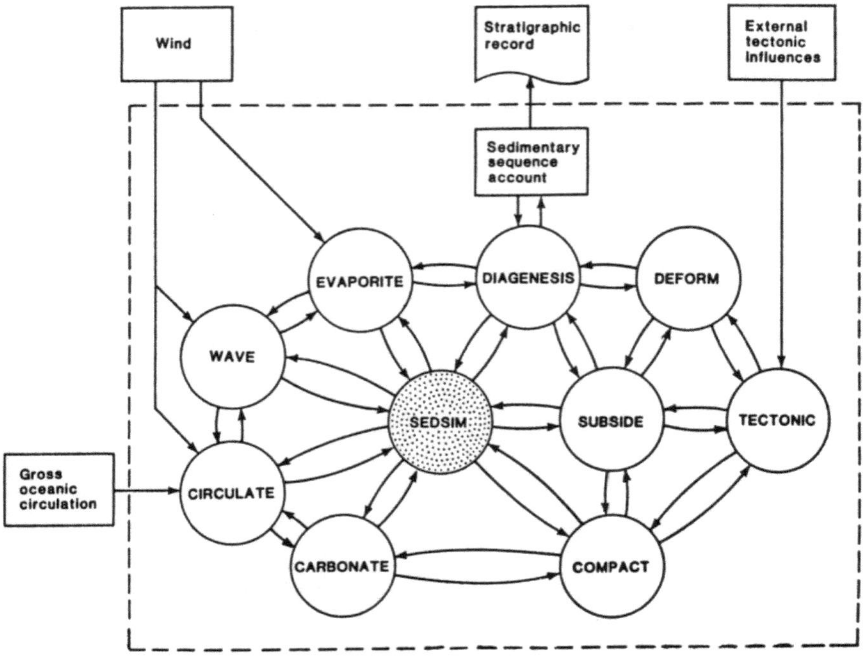

Figure 1.1 Modules proposed for linkage with SEDSIM that deal with most major processes that create sedimentary basins. Names of modules connote their respective roles.

The three versions represent SEDSIM's progressive evolution from a flow-only model, to a rudimentary model capable of dealing with only a single type of clastic sediment, and finally culminating in an advanced model capable of dealing with different types of clastic sediment. Each represents an extension of the previous one, but each can be used independently or linked with modules shown in Figure 1.1.

One of our purposes in this monograph is to advance the "systems viewpoint" in geology by providing simulation procedures that can be used experimentally and that represent sedimentary basins as complex dynamic systems. Experiments on sedimentary basins cannot be performed with small-scale physical models, but experiments are feasible with dynamic simulation models run on a computer. This monograph describes the components involved in the creation and transformation of sedimentary basins that consist of the erosion, transport, and deposition of clastic sediment by running water. Flow and transport are, of course, important components of the overall rock cycle, and hopefully SEDSIM will prove to be a major step in developing procedures that can simulate processes that shape and transform the earth's crust as a whole.

SEDSIM'S COMPUTER PROGRAMS

The three versions of SEDSIM are written in standard FORTRAN-77. As such, they are relatively "transportable" from computer to computer, but they incorporate subroutine calls that are specific for Tektronix 4115 high-resolution color terminals, and in that sense they are linked with specific hardware devices. However, adroit programmers can adapt SEDSIM's modules to other graphics equipment.

SEDSIM's versions are modular in form, employing short main programs that call subroutines to accomplish specific tasks. The aggregate number of lines of FORTRAN code for each of SEDSIM's versions is moderate (SEDSIM3 has about 2100 lines, Appendixes B and C), but recursion is high, resulting in long run times. Simulations are usually run in batch mode, with results written onto disk files. During operation, both grid parameters and fluid-element parameters are updated in cycles representing time increments of a few seconds duration. Consistent use of such short time increments would make calculations involving intervals of geologic time prohibitive because of the vast amount of arithmetic. Fortunately, simple procedures can be employed to reduce the amount of arithmetic.

Input to SEDSIM involves auxiliary computer programs that interactively generate data files containing an initial topographic grid as well as the parameters that regulate fluids and sediments. Output is handled by another interactive program that prepares disk files for display as maps, perspective diagrams, and vertical sections. The graphics procedures are relatively fast and easy to use, yielding maps and vertical sections that can be stored and subsequently shown in quick succession on a color terminal, representing the progressive evolution of a sedimentary basin through geologic time. If the intervals between individual displays are short, the effect of motion is conveyed, much as in animated video form.

CLASSIFYING SIMULATION MODELS

Simulation as employed here involves creating dynamic mathematical models of actual systems and then performing experiments that imitate the actual systems. Although simulation models have been devised in many fields and are represented by a vast literature, simulation of geologic processes is in relative infancy and has been extensively applied only in hydrology, in management of oil and gas reservoirs, and, to a lesser extent, in geomorphology.

Table 1.1 classifies simulation models according to whether they are physical or mathematical, random or deterministic, nondimensional, one-, two-, or three-dimensional, and static or dynamic. Table 1.1 ranks SEDSIM's principal features within such a spectrum of simulation models.

TABLE 1.1 Classification of Simulation Models Listing Representative Earth Science Examples, including SEDSIM

Criterion	Categories		Gravity and magnetic response of ore bodies	Artificial seismic sections	Stratigraphic column Markov chains	Geostatistical properties of ore deposits	Causes of glaciations	Geochemical cycle of elements	Oil-reservoir and groundwater flow	Steady-state potential flow	Flood routing and environmental engineering	Beach processes	Sedimentary basin sections	Laboratory experiments in sedimentation	SEDSIM
Method	Physical													x	
	Mathematical	Analytic	x				x	x							
		Computer	x	x	x	x	x	x	x	x	x	x	x		X
Operating mechanism	Probabilistic				x	x									
	Deterministic	Regular	x	x			x	x	x	x	x		x		
		Chaotic										x		x	X
Spatial dimensions	None						x	x							
	1-D				x										
	2-D	Section		x		x			x	x			x		
		Plan	x			x			x	x	x	x			
	3-D					x			x	x				x	X
Time	Static		x	x	x	x				x					
	Dynamic (process modeling)						x	x	x		x	x	x	x	X

If motion or change is represented, then time is implicitly represented. Time and motion are interdependent, for neither can be defined in the absence of the other. The "real" world thus occupies at least four dimensions: three of space and one of time. The spatial variables can be placed in discrete form, represented by an aggregate of blocks or cells or by a gridwork of points in space. For example, variables that represent sedimentary facies as proportions of sand, silt, and clay can be approximated by values assigned to a three-dimensional grid of points. If the model is dynamic, the fourth dimension of time also can be represented as a succession of such three-dimensional grids, each separated by a specific increment of time.

Dynamic models are important in geology because they can recreate events that happened in the geologic past. For example, processes that created ancient evaporite deposits can be represented by dynamic models that allow basin topography, circulation patterns, inflow rates, and evaporation rates to be adjusted as experiments are performed.

Broadly, sedimentation simulation models can be divided into (1) physical models, such as flumes or sedimentation tanks, that are scale models of actual physical processes, and (2) mathematical models, where relationships are expressed as equations and logic operations, generally in the form of computer programs.

Physical models directly represent actual systems, but generally they are hindered because some variables cannot be scaled downward or upward. For example, we might attempt to scale down the Mississippi river's delta to 1/10,000 of its actual linear dimensions, but we cannot proportionally scale down the Mississippi's fluid viscosity or density, nor can we scale down the physical properties of the particles of sediment that the Mississippi transports. But even if we could scale these factors down, fluid flow and sediment transport operate differently at much reduced scales(Mathematical models, on the other hand, do not suffer from those scaling limitations, but mathematical models do suffer from the simplifications required if the models are to be feasible for computation. For example, the necessity of dividing area or space into cells and time into steps poses severe limitations. There are other issues to be considered also, including boundaries, number of dimensions, and the magnitudes of time and change. These general issues are discussed in an overview by Harbaugh and Bonham-Carter (1970).

Simulation models may contain elements that are deliberately random when cause-and-effect relationships are subjected to influences governed by probability distributions. For example, one-dimensional sequences of sedimentary beds can be simulated by random draws governed by Markov transition matrices. yielding beds whose thickness or composition is regulated statistically by the transition matrices. Alternatively, simulation models can be totally deterministic, so each state of the model leads unequivocally to the next. However, even totally deterministic models may seem to behave randomly, particularly when small differences between initial states result in large differences in subsequent states. SEDSIM is actually deterministic, but it operates with apparently random behavior. For example, when SEDSIM2 or SEDSIM3 simulates a braided stream, individual experiments may begin with straight channels, but minute differences in initial flow conditions may bring about large differences in the sinuous channels formed subsequently. Of course, actual streams also behave unpredictably, even though they are governed by deterministic physical laws, so small variations may lead to large subsequent variations. But SEDSIM's versions contain no deliberately random elements, and their behavior in specific experiments can be repeated precisely if initial conditions are unchanged. However, without actually performing a simulation experiment,

SEDSIM's response can be predicted only in a very generalized way, and the details of the response are not predictable except by operating the model.

Because of SEDSIM's behavior, we must be cautious about the results of a single experiment. For example, small changes in initial conditions when turbidity flows are simulated may cause subsequent flows to follow different paths as they move over depositional fans that they create. If we conducted only a single such experiment, we might conclude falsely that deposits necessarily form channels on a particular side of a fan because they formed there during the experiment. Conclusions about features produced in experiments should involve successions of simulations in which moderate changes in initial conditions have been made.

Simulation models can be classified according to the number of dimensions that they represent. Some are dimensionless, such as those dealing with geochemical cycles in which neither area nor space is represented. One-dimensional models involving sequences of events were once popular, such as bedding in sedimentary sequences, but two-dimensional models are now more common, and full three-dimensional models are feasible. SEDSIM's versions are two-dimensional in certain aspects and three-dimensional in others, and might be classified as "two-plus" or "quasi-three-dimensional." All three versions of SEDSIM simulate flow in two horizontal dimensions, but depth of flow is considered, and therefore SEDSIM represents flow in more than two dimensions but not completely in three. On the other hand, SEDSIM2 and SEDSIM3 are fully three-dimensional in their representation of sediment that has been deposited.

SEDSIM is flexible in that flow and sedimentation can be simulated over wide ranges of conditions for relatively long intervals of time. There are, however, limitations imposed (1) by the need to control the simulation, (2) by irreversibility, (3) by boundary conditions, (4) by simplification with respect to the actual system represented, and (5) by computing power.

CONTROLLING THE SIMULATIONS

Even if an immense physical laboratory were available in which sedimentation experiments could be conducted at full scale, with whole sedimentary basins being represented, we would still be limited in our attempts to recreate past details of sedimentary basins because the conditions under which the basins formed can never be fully known. Accurate reconstruction of an ancient sedimentary basin would require detailed information about its former topography, including the locations of rivers that flowed into it, and their flow rates and volumes, and the composition of sediment that they carried. We would also need to know details of the basin's structural evolution. Therefore, limitations stemming from lack of knowledge about factors that controlled ancient sedimentary basins cannot be resolved directly by SEDSIM. However, we should be able to establish ranges for past depositional and structural conditions by performing repeated simulation experiments and comparing the results with features preserved in actual basins. Of course, different depositional and structural assumptions may yield similar responses in simulation experiments, but we should be able to constrain the parameters that control experiments so that reasonable assumptions eventually can be selected and realistic experimental results obtained. Realistic performance is a major goal in process simulation.

IRREVERSIBILITY AND TIME'S ARROW

We have often been asked whether SEDSIM could be run "backwards" through time. It might seem ideal to have a simulation model that would run backwards and reconstruct the past from information about the present. Unfortunately, few process models will run backwards, and SEDSIM definitely will not. Creating a process model that would run backwards might be likened to the challenge of reconstructing a person's life history from his or her present activities. Such reconstructions are difficult and involve immense uncertainty. Furthermore, in geology we are dealing with processes that are inherently irreversible.

It is true that equations that govern physical phenomena can be "operated" in reverse. For example, the frictionless single-fluid element experiment described in Chapter 3 can be run backwards, so the fluid element retraces its path exactly. There is no uncertainty; the fluid element returns precisely to its initial position. But, suppose that friction is introduced. The model can still be reversed, but it becomes increasingly difficult to retrace the path of the fluid element. As time advances, information is progressively lost, and eventually it becomes impossible to reverse the model. At that point we can operate the model only in the forward direction. "Time's arrow" therefore restricts the model's operation to the forward direction, and time's arrow represents the progressive loss of information about the past, thereby decreeing that the model cannot be reversed.

If we test assumptions about the past with a process model that operates only in the "forward" direction, we progressively lose information and thereby create increasing uncertainty about the past, regardless of the "correctness" of the initial conditions. Although we can trace the consequences of different initial assumptions, equally plausible results may ensue from different initial assumptions. Thus, even though it is impossible to reverse a process model such as SEDSIM, the model's performance may enable us to assess the reasonableness of the original assumptions. Therefore, one of process simulation's main roles is to enable us to deduce the consequences that stem from different assumptions while consistently operating forward through time. Thus, simulation may not resolve alternative hypotheses, but it can serve us well by helping us to deduce the consequences of our assumptions.

Philosophical problems in interpreting the geological past with models that work only forward are as old as the geological sciences themselves. Geologists also engage in forward modeling when they deduce the consequences of assumptions incorporated in conceptual models of geologic processes. SEDSIM's strength is that it provides disciplined, quantitative procedures for deducing the consequences of assumptions involving the erosion, transport, and deposition of clastic sediment.

BOUNDARY CONDITIONS

Boundaries are needed because we cannot simulate a large actual system as a whole. The conditions established at boundaries are also part of the control parameters. In geologic simulation, we might avoid problems at boundaries if we could represent the earth as a whole. Unfortunately, the whole earth is much too large to represent in a simulation model. An entire sedimentary basin also may be much too large because it is part of a still larger system and therefore must be separated by boundaries. Thus, geographic boundaries must be established. Physical models, such as flumes and sand tables, also have boundaries, and also must incorporate assumptions about how the simulated system is to be isolated from its surroundings. The treatment at boundaries thus is critical, both in physical and in mathematical models.

SEDSIM incorporates assumptions about how processes operate adjacent to boundaries. Flow is allowed to enter or exit freely, but erosion and deposition are restricted at boundaries, with most of the immediate boundary effects confined to narrow zones adjacent to boundaries. However, boundary conditions may affect the entire area of the simulation. For example, "depositional equilibrium" may be attained within the area of a sedimentation model so that the amount of sediment that enters during an interval of time is equal to that which leaves. However, if the simulated area is enlarged by shifting the boundaries outward, deposits will continue to form, and their gross thickness and other properties may differ once equilibrium is again established. Thus, changes at boundaries generally affect the entire system being simulated.

SEDSIM provides for inflows of fluid and sediment at a number of fixed points, represented as "sources," that are usually located near the edges of the simulated area. These inflows are also boundary conditions, and they represent "exogenous" inputs from the "outside" world. Inflows at sources may be far from equilibrium with respect to flow conditions within the system. When flow, slope, and sediment load are not in equilibrium, unrealistic topographic mounds or depressions may be created locally. Only after the flows have moved sufficiently far from the sources, will equilibrium have been attained and will deposition be realistic. Thus, the effects of inflows of fluid and sediment at boundaries may be propagated over the entire system.

SIMPLIFICATIONS WITH RESPECT TO ACTUAL SYSTEMS

Dynamic process models such as SEDSIM are necessarily simplified versions of actual systems. Hopefully, the simplifications do not result in unduly unrealistic performance, but some loss of "reality" is inevitable. For example, SEDSIM uses only two horizontal dimensions to represent flow, a simplification that reduces computing effort but also decreases realism. One consequence is that sedimentary features affected by vertical movement within flows, such as ripples whose height is small with respect to flow depth, cannot be produced directly.

The sediment-transport equations used by SEDSIM represent another simplification. If we used different equations for different conditions of flow, we would achieve more realistic results, but specific transport equations have not been devised that are appropriate for all conditions, and therefore we must select among alternatives.

Discretizations of models by representing time in steps, and area or space with cellular gridworks, are also simplifications. We lose resolution, of course, when we use discrete cells and discrete time steps, but we can reduce the loss of resolution by making the cells smaller and the time steps shorter. But, such improvements are accompanied by increases in computing effort, posing complex trade-offs in the search for an optimum procedure.

COMPUTING POWER

SEDSIM is limited by available computer power and computer time. Most experiments described here were devised with economy of computing effort in mind. For example, multiple-sediment experiments with SEDSIM3 involve only about 5000 fluid elements, a very small number, considering that the number of fluid elements is virtually infinite in any actual flow. Furthermore, experiments described here involve geographic representation with horizontal grids that do not exceed a maximum of 31 columns and 31 rows (961 cells) and a three-dimensional representation of sedimentary deposits by grids that do not exceed 31 rows, 31 columns, and 800 layers (768,800 cells). These grids are relatively coarse and use few fluid elements; collectively they require only about $\frac{1}{2}$ megabyte (M byte) of core memory, which is modest, considering that most midsize computers have at least several megabytes of memory. Thus, computing constraints are not generally imposed by the size of the computer memories, and instead the limits on numbers of grid cells and fluid elements are dictated by computer speed.

Computing speed is thus the major constraint. The longest experiments described here involve use of SEDSIM3 to simulate deposition of large deltas for

50,000 years of geologic time. These experiments require run times of four to eight hours or longer on computers operating at 12 million instructions per second. Computer time is a major consideration. In fact, most personal computers have enough memory to run similar experiments, but they are so slow that days or weeks would be required for the calculations. As faster computers are developed, SEDSIM's usefulness should increase, particularly when used with computers that employ parallel processors to perform multiple operations simultaneously. Furthermore, such increases in speed could be accompanied by graphic output during experiments, so experiments could be controlled while in progress by changing the control parameters.

CHAPTER 2

Adapting the Navier-Stokes Equations for a Flow Model

Some sedimentation simulation models do not represent flow (they represent only the transport of sediment), or they may represent only steady flow in open channels of constant cross section. SEDSIM, however, can represent unsteady flow in irregular channels, a capability based on adaptations of the Navier-Stokes equations. The Navier-Stokes equations in their original form completely describe flow in three dimensions. Unfortunately, they cannot be incorporated directly into computer programs because they are in differential form. However, after some simplifying assumptions, they can be represented as numerical approximations and incorporated in computer programs. The simplified Navier-Stokes equations then can represent flow in channels whose cross sections are irregular and in channels that merge as tributaries or diverge as distributaries. The simplified equations contain two parameters that depend partly on flow conditions, such as channel shape and flow velocity. These parameters must be calibrated with semiempirical formulas.

The most important simplification in our use of the Navier-Stokes equations involves the assumption that the velocity of flow does not vary with depth at a given moment. This assumption may seem unwarranted, because actual flow velocities vary with depth. But if the flow equations are linked with appropriate equations for sediment load, the representation of sediment transport is suitably realistic. We could represent flow velocities that vary with depth, but the computational effort would increase greatly.

The Navier-Stokes equations were adapted initially in SEDSIM1, and were subsequently linked in SEDSIM2 and SEDSIM3 with other equations that govern erosion, transport, and deposition of sediment. SEDSIM1 thus provides the framework of flow equations that all three versions of SEDSIM utilize.

EULERIAN VERSUS LAGRANGIAN REPRESENTATION

Flow equations can have two distinct representations, Eulerian and Lagrangian, each of which requires different numerical procedures. In Eulerian representation (Fig. 2.1*a*), flow is described by velocity, acceleration, and density at points fixed in space. Eulerian methods are convenient when velocity does not change at the fixed points, and the flow equations can be solved by finite-difference or finite-element methods. Unfortunately, Eulerian methods are not suited for unsteady flow and therefore are not useful for SEDSIM, which incorporates unsteady flow.

Lagrangian representations do not employ a fixed grid, and therefore neither finite-difference or finite-element methods can be used. Instead, the flow parameters can be represented with respect to the fluid itself, providing a moving frame of reference (Fig. 2.1*b*). A difficulty arises, however, because numerical approximations of flow equations require reference to points that move continuously with respect to each other, and deformable grids are not practical. The difficulty can be circumvented by utilizing many fluid elements whose positions and velocities are recalculated at each time step.

Another disadvantage of Eulerian representation is that suspended sediment and other substances transported in the simulated flow may "diffuse," even though diffusion is not intended. The diffusion arises artificially because the computing procedure causes transported sediment to be repeatedly apportioned between adjacent grid cells, thus paralleling actual diffusion processes. Use of Lagrangian procedures eliminates the unintended diffusion.

MARKER-IN-CELL REPRESENTATION

Given the merits of both Lagrangian and Eulerian procedures, there is a procedure that combines them both (Fig. 2.1*c*), called the *marker-in-cell* method (Harlow, 1964), or *particle-cell* method (Hockney and Eastwood, 1981). Marker-in-cell representation can simulate free-surface flow in a vertical cross section (Harlow, 1964), as might be used to represent flows when a dam collapses or when a sluice gate is opened. Marker-in-cell representation is also useful for simulating flow in two horizontal dimensions (Cheng, 1983). Full three-dimensional marker-in-cell procedures are computationally exhausting and have been mostly applied in particle physics (Buneman et al., 1980), but with enough computing power they would be useful for simulating sediment transport.

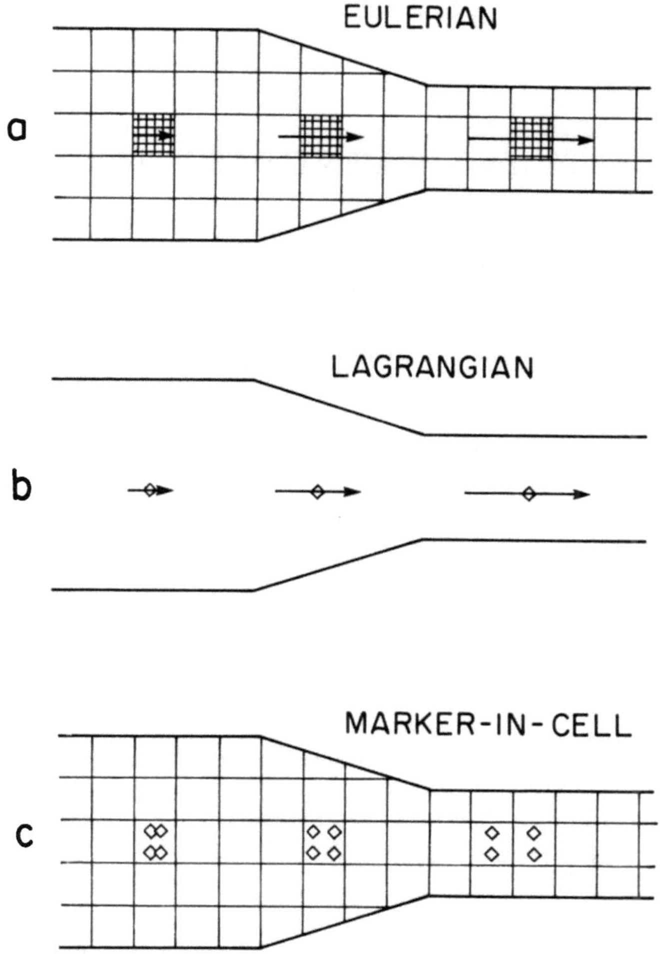

Figure 2.1 Plan view of hypothetical channel schematically illustrating three procedures for numerical representation of flow. (*a*) Eulerian procedure utilizes fixed grid and is effective for steady flow. (*b*) Lagrangian procedure refers flow parameters to moving fluid elements. (*c*) Marker-in-cell procedure uses "markers" or fluid elements to represent unsteady flow and sediment transport, but retains fixed grid for some flow parameters to facilitate computation.

SEDSIM uses a marker-in-cell technique in two horizontal dimensions. Flow velocity and sediment load are represented at points that move with the fluid (Fig. 2.1c). A two-dimensional square grid is used to represent depth of flow, as well as elevation of the sea bottom or water-sediment interface. Thus SEDSIM is "two-plus" dimensional in its representation of flow, requiring use of depth-mean parameters of flow that represent an average of flow properties along a vertical line extending through the flow. The resulting formulation is suitable for simulating free-surface flow under a variety of different flow conditions.

CONTINUITY AND MOMENTUM EQUATIONS

The Navier-Stokes equations combine the continuity equation (Eq. 2.1) and the momentum equation (Eq. 2.2) to provide a complete mathematical description of flow for an isotropic Newtonian fluid. An isotropic Newtonian fluid is a fluid whose properties are uniform in all directions and that follows Newton's laws of motion. The continuity equation incorporates the conservation of mass:

$$\frac{\partial \rho}{\partial t} + \nabla \cdot \rho \mathbf{q} = 0 \tag{2.1}$$

where

ρ = fluid density
t = time
$\mathbf{q}$ = flow velocity vector

whereas the momentum equation equates changes in momentum by the fluid:

$$\rho\left(\frac{\partial \mathbf{q}}{\partial t} + (\mathbf{q} \cdot \nabla)\mathbf{q}\right) = -\nabla p + \nabla \cdot \mu U + \rho(\mathbf{g} + \Omega \mathbf{q}) \tag{2.2}$$

where

p = pressure
μ = fluid viscosity
U = Navier-Stokes tensor

$$= \begin{bmatrix} 2\frac{\partial u}{\partial x} - \frac{2}{3}\nabla \cdot \mathbf{q} & \frac{\partial u}{\partial y} + \frac{\partial v}{\partial x} & \frac{\partial u}{\partial z} + \frac{\partial w}{\partial x} \\ \frac{\partial u}{\partial y} + \frac{\partial v}{\partial x} & 2\frac{\partial v}{\partial y} - \frac{2}{3}\nabla \cdot \mathbf{q} & \frac{\partial v}{\partial z} + \frac{\partial w}{\partial y} \\ \frac{\partial u}{\partial z} + \frac{\partial w}{\partial x} & \frac{\partial v}{\partial z} + \frac{\partial w}{\partial y} & 2\frac{\partial w}{\partial z} - \frac{2}{3}\nabla \cdot \mathbf{q} \end{bmatrix}$$

u, v, w = components of flow velocity $\mathbf{q}$ in the direction of x, y, and z axes, respectively

Ω = Coriolis tensor

$$= \begin{bmatrix} 0 & 2\omega \sin\phi & -2\omega\sin\phi \\ -2\omega\sin\phi & 0 & 0 \\ -2\omega\cos\phi & 0 & 0 \end{bmatrix}$$

ω = angular rotational velocity of the earth

ϕ = latitude

$\mathbf{g}$ = gravitational acceleration vector

If the fluid is homogeneous, incompressible, and remains at constant temperature, the density ρ and viscosity μ can be considered constant. The Coriolis acceleration $\Omega\mathbf{q}$ is so small in most of SEDSIM's applications that it can be ignored. Marine currents on an oceanic scale are strongly affected by the Coriolis acceleration, and SEDSIM cannot simulate large-scale circulation patterns because the Coriolis acceleration is not incorporated.

If ρ and μ are constant, and $\Omega\mathbf{q}$ equals 0, the system of Equations 2.1 and 2.2 becomes

$$\nabla \cdot \mathbf{q} = 0 \tag{2.3}$$

$$\frac{\partial \mathbf{q}}{\partial t} + (\mathbf{q} \cdot \nabla)\mathbf{q} = -\nabla\Phi + \nu\nabla^2\mathbf{q} + \mathbf{g} \tag{2.4}$$

where

Φ = ratio of pressure to constant density = p/ρ

ν = kinematic viscosity = μ/ρ

The solution to the system formed by Equations 2.3 and 2.4 is completely determined when initial conditions and boundary conditions are specified. Initial conditions involve flow at time 0:

$$\mathbf{q}(x, y, z, t_0) = \mathbf{q}_0 \tag{2.5}$$

BOUNDARY CONDITIONS FOR FREE-SURFACE FLOW

Boundary conditions must be specified at the interface between the fluid and its surroundings. This interface is assumed to be either a free surface (i.e., water in contact with air) or a rigid impermeable surface, such as the sea bottom or the bed of a stream. By "rigid" we mean that the rate of change of elevation of the sea bottom or stream bed due to erosion or deposition is negligibly small compared with the rates of fluid movement.

A free surface can be described by the implicit equation

$$f(x, y, z, t) = 0 \tag{2.6}$$

The kinematic equation of the free surface is

$$u(H)\frac{\partial f}{\partial x} + v(H)\frac{\partial f}{\partial y} + w(H)\frac{\partial f}{\partial z} + \frac{\partial f}{\partial t} = \frac{Df}{Dt} = 0 \tag{2.7}$$

where H = free surface elevation (with respect to sea level).

The rigid impermeable surface is assumed to be a function of x and y in which every point in the xy plane has only one elevation, thus prohibiting representation of vertical or overhanging slopes. Furthermore, we assume that the elevation of the water surface is also a function of x and y, and that the flow is bound below by a rigid impermeable surface and above by a free surface. Thus, an ordinary stream flows over the stream bed (the rigid impermeable surface) and is bounded above by the air-water interface (the free surface). Equation 2.6 then can be expressed as an explicit equation of z:

$$z = H(x, y, t) \tag{2.8}$$

where

z = vertical coordinate (above sea level)
H = elevation of water surface

Now Equation 2.7 yields

$$\frac{\partial H}{\partial t} = w(H) - u(H)\frac{\partial H}{\partial x} - v(H)\frac{\partial H}{\partial v} \tag{2.9}$$

Equation 2.9 defines the boundary conditions for the free surface of the flow.

At the rigid impermeable boundary we assume that flow perpendicular to the boundary is zero, permitting further simplification of the flow equations. Thus,

$$\mathbf{q} \cdot \mathbf{N} = 0 \tag{2.10}$$

where **N** = any vector perpendicular to the boundary. In particular,

$$\mathbf{N} = \left(-\frac{\partial Z}{\partial x}, -\frac{\partial Z}{\partial y}, 1 \right) \tag{2.11}$$

where Z = topographic elevation (with respect to sea level). Therefore,

$$-u(Z)\frac{\partial Z}{\partial x} - v(Z)\frac{\partial Z}{\partial y} + w(Z) = 0 \tag{2.12}$$

$$w(Z) = u(Z)\frac{\partial Z}{\partial x} + v(Z)\frac{\partial Z}{\partial y} \tag{2.13}$$

Equation 2.13 defines the boundary condition for the rigid impermeable surface of the flow.

TWO-DIMENSIONAL CONTINUITY EQUATION

We can adapt the simplified Navier-Stokes continuity equation 2.3 to obtain a two-dimensional continuity equation that involves horizontal flow velocities and also represents the depth of fluid confined between a rigid boundary below and a free surface above. From Equation 2.3, we have

$$\frac{\partial w}{\partial z} = -\frac{\partial u}{\partial x} - \frac{\partial v}{\partial y} \tag{2.14}$$

Integrating vertically over the depth of the flow gives

$$\int_Z^H \frac{\partial w}{\partial z}\,dz = \int_Z^H \left(-\frac{\partial u}{\partial x} - \frac{\partial v}{\partial y} \right) dz \tag{2.15}$$

$$w(H) - w(Z) = \int_Z^H -\frac{\partial u}{\partial x}\,dz + \int_Z^H -\frac{\partial v}{\partial y}\,dz \tag{2.16}$$

$$w(H) = \int_Z^H -\frac{\partial u}{\partial x}\,dz + \int_Z^H -\frac{\partial v}{\partial y}\,dz + w(Z) \tag{2.17}$$

Both H and Z are functions of x and y (the horizontal coordinates). Therefore, applying Leibnitz's rule, we obtain

$$w(H) = -\frac{\partial\left(\int_Z^H u\,dz\right)}{\partial x} + u(H)\frac{\partial H}{\partial x} - u(Z)\frac{\partial Z}{\partial x}$$
$$-\frac{\partial\left(\int_Z^H v\,dz\right)}{\partial y} + v(H)\frac{\partial H}{\partial y} - v(Z)\frac{\partial Z}{\partial y} + w(Z) \qquad (2.18)$$

It is convenient at this point to define the vertically averaged velocities as

$$\bar{u} = \frac{1}{H - Z}\int_Z^H u\,dz \qquad (2.19)$$

$$\bar{v} = \frac{1}{H - Z}\int_Z^H v\,dz \qquad (2.20)$$

and

$$\mathbf{Q} = \bar{u}\mathbf{i} + \bar{v}\mathbf{j} \qquad (2.21)$$

where

$\mathbf{Q}$ = vertically averaged horizontal velocity (could also be noted as $\bar{\mathbf{q}}_{xy}$)
$\mathbf{i}, \mathbf{j}$ = vectors of unit length in the direction of x and y axes, respectively

Then Equation 2.18 becomes

$$w(H) = -\frac{\partial(H - Z)\bar{u}}{\partial x} + u(H)\frac{\partial H}{\partial x} - u(Z)\frac{\partial Z}{\partial x}$$
$$-\frac{\partial(H - Z)\bar{v}}{\partial y} + v(H)\frac{\partial H}{\partial y} - v(Z)\frac{\partial Z}{\partial y} + w(Z) \qquad (2.22)$$

Using Equation 2.13 to replace $w(Z)$ and canceling opposite terms, we get

$$w(H) = -\frac{\partial(H - Z)\bar{u}}{\partial x} + u(H)\frac{\partial H}{\partial x} - \frac{\partial(H - Z)\bar{v}}{\partial y} + v(H)\frac{\partial H}{\partial Z} \qquad (2.23)$$

Now Equation 2.23 can be used to replace $w(H)$ in Equation 2.9, which, after canceling the opposite terms, yields

$$\frac{\partial h}{\partial t} = \frac{\partial H}{\partial t} = -\frac{\partial((H - Z)u)}{\partial x} - \frac{\partial((H - Z)v)}{\partial y} = -\nabla \cdot ((H - Z)\mathbf{Q}) \tag{2.24}$$

Thus,

$$\frac{\partial h}{\partial t} = \frac{\partial H}{\partial t} = -\nabla \cdot (h\mathbf{Q}) \tag{2.25}$$

where h = fluid depth = $H - Z$.

Equation 2.25 is the continuity equation for the depth-mean representation. It says that if the fluid diverges, its surface must drop, and vice versa. Divergence is a measure of the rate at which flow lines separate within a field of flow. If positive, the lines diverge; if negative, the lines converge. Equation 2.25 constitutes the first of two main equations in the simplified two-dimensional flow model.

TWO-DIMENSIONAL MOMENTUM EQUATION

The second equation in the flow model is derived from Equation 2.4, which can be rewritten to represent components that lie along the x, y, and z axes, respectively, as follows:

For the x component:

$$\frac{\partial u}{\partial t} + u\frac{\partial u}{\partial x} + v\frac{\partial u}{\partial y} + w\frac{\partial u}{\partial z} = -\frac{1}{\rho}\frac{\partial p}{\partial x} + \frac{\mu}{\rho}\left(\frac{\partial^2 u}{\partial x^2} + \frac{\partial^2 u}{\partial y^2} + \frac{\partial^2 u}{\partial z^2}\right) \tag{2.26}$$

For the y component:

$$\frac{\partial v}{\partial t} + u\frac{\partial v}{\partial x} + v\frac{\partial v}{\partial y} + w\frac{\partial v}{\partial z} = -\frac{1}{\rho}\frac{\partial p}{\partial y} + \frac{\mu}{\rho}\left(\frac{\partial^2 v}{\partial x^2} + \frac{\partial^2 v}{\partial y^2} + \frac{\partial^2 v}{\partial z^2}\right) \tag{2.27}$$

For the z (vertical) component:

$$\frac{\partial w}{\partial t} + u\frac{\partial w}{\partial x} + v\frac{\partial w}{\partial y} + w\frac{\partial w}{\partial z} = -\frac{1}{\rho}\frac{\partial p}{\partial z} + \frac{\mu}{\rho}\left(\frac{\partial^2 w}{\partial x^2} + \frac{\partial^2 w}{\partial y^2} + \frac{\partial^2 w}{\partial z^2}\right) + g \qquad (2.28)$$

where $g = |\mathbf{g}|$.

Equation 2.28 is not needed if the distribution of fluid pressures is considered to be hydrostatic. Equations 2.26 and 2.27 can be integrated with respect to z, respectively yielding

$$\underbrace{\int_Z^H \frac{\partial u}{\partial t}\,dz}_{\text{Ⓐ}} + \underbrace{\int_Z^H \left(u\frac{\partial u}{\partial x} + v\frac{\partial u}{\partial y}\right) dz}_{\text{Ⓑ}} + \underbrace{\int_Z^H w\frac{\partial u}{\partial z}\,dz}_{\text{Ⓒ}} = \underbrace{\int_Z^H -\frac{1}{\rho}\frac{\partial p}{\partial x}\,dz}_{\text{Ⓓ}} + \underbrace{\int_Z^H \frac{\mu}{\rho}\left(\frac{\partial^2 u}{\partial x^2} + \frac{\partial^2 u}{\partial y^2}\right) dz}_{\text{Ⓔ}} + \underbrace{\int_Z^H \frac{\mu}{\rho}\frac{\partial^2 u}{\partial z^2}\,dz}_{\text{Ⓕ}} \qquad (2.29)$$

and

$$\int_Z^H \frac{\partial v}{\partial t}\,dz + \int_Z^H \left(u\frac{\partial v}{\partial x} + v\frac{\partial v}{\partial y}\right) dz + \int_Z^H w\frac{\partial v}{\partial z}\,dz = \int_Z^H -\frac{1}{\rho}\frac{\partial p}{\partial x}\,dz + \int_Z^H \frac{\mu}{\rho}\left(\frac{\partial^2 v}{\partial x^2} + \frac{\partial^2 v}{\partial y^2}\right) dz + \int_Z^H \frac{\mu}{\rho}\frac{\partial^2 v}{\partial z^2}\,dz \qquad (2.30)$$

Equations 2.29 and 2.30 are thus the two-dimensional momentum equations, equating horizontal forces and changes in momentum for an element of fluid.

INTEGRATING THE MOMENTUM EQUATION

Each term in Equation 2.29 has been labeled (encircled letters) so that integration may be carried out separately for each term. The procedure for integrating Equation 2.30 is not shown in detail, but it is similar to the procedure described for Equation 2.29. The combination of both equations ultimately results in Equation 2.78. To integrate the terms of Equation 2.29 along the vertical dimension, we must assume that the vertical flow-velocity profile is the same everywhere, because otherwise the system represented by Equations 2.26–2.28 cannot be simplified effectively. Thus, if the flow-velocity profile has the same shape everywhere in the flow, then there is a function r (Fig. 2.2) such that, for every point (x_0, y_0),

$$\mathbf{q}_{xy}(x_0, y_0, z) = r\left(\frac{z - Z}{H - Z}\right)\mathbf{Q}(x_0, y_0) \tag{2.31}$$

where

$\mathbf{q}_{xy}$ = horizontal velocity = $u\mathbf{i} + v\mathbf{j}$
z = vertical coordinate with respect to sea level (within the flow $H \geq z \geq Z$)

Let

$$s = \frac{z - Z}{H - Z} \tag{2.32}$$

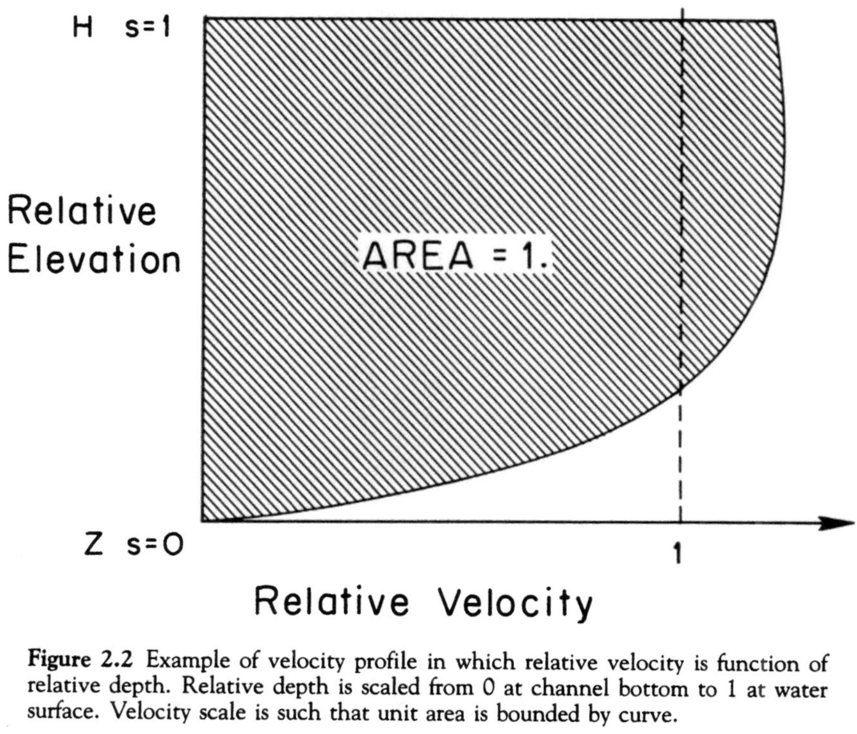

Figure 2.2 Example of velocity profile in which relative velocity is function of relative depth. Relative depth is scaled from 0 at channel bottom to 1 at water surface. Velocity scale is such that unit area is bounded by curve.

It follows that

$$\int_0^1 r(s)\, ds = 1 \tag{2.33}$$

The function r is the velocity profile.

Integration of Equation 2.29 can then proceed as follows:

Integration of Ⓐ:

$$Ⓐ = \int_Z^H \frac{\partial u}{\partial t}\, dz \tag{2.34}$$

$$Ⓐ = \frac{\partial \int_Z^H u\, dz}{\partial t} - u(H)\frac{\partial H}{\partial t} + u(Z)\frac{\partial z}{\partial t} \tag{2.35}$$

Assuming that the change in topography with time is very small, we can write

$$Ⓐ = \frac{\partial((H - Z)\bar{u})}{\partial t} - u(H)\frac{\partial H}{\partial t} \tag{2.36}$$

$$Ⓐ = (H - Z)\frac{\partial \bar{u}}{\partial t} + \frac{\partial H}{\partial t}\bar{u} - \frac{\partial H}{\partial t}u(H) \tag{2.37}$$

$$Ⓐ = (H - Z)\frac{\partial \bar{u}}{\partial t} + \frac{\partial H}{\partial t}\bar{u}(1 - r(1)) \tag{2.38}$$

Let

$$\alpha = (1 - r(1)) \tag{2.39}$$

Then

$$Ⓐ = h\frac{\partial \bar{u}}{\partial t} + \frac{\partial H}{\partial t}\bar{u}\alpha \tag{2.40}$$

Integration of Ⓑ:

$$Ⓑ = \int_Z^H \left(u\frac{\partial u}{\partial x} + v\frac{\partial u}{\partial y}\right) dz \tag{2.41}$$

$$Ⓑ = \underbrace{\int_Z^H u\frac{\partial u}{\partial x}\, dz}_{\text{B1}} + \underbrace{\int_Z^H v\frac{\partial u}{\partial y}\, dz}_{\text{B2}} \tag{2.42}$$

$$\textcircled{B1} = \int_Z^H u\frac{\partial u}{\partial x}\,dz = (H - Z)\int_0^1 r(s)\overline{u}r(s)\frac{\partial \overline{u}}{\partial x}\,ds \tag{2.43}$$

$$\textcircled{B1} = (H - Z)\overline{u}\frac{\partial \overline{u}}{\partial x}\int_0^1 r^2(s)\,ds \tag{2.44}$$

Let

$$\beta = \int_0^1 r^2(s)\,ds \tag{2.45}$$

Then, after integrating $\textcircled{B2}$ (Eq. 2.42) in similar fashion, we obtain

$$\textcircled{B} = h(\mathbf{Q}\cdot\nabla)\overline{u}\beta \tag{2.46}$$

Integration of $\textcircled{C}$:

$$\textcircled{C} = \int_Z^H w\frac{\partial u}{\partial z}dz \tag{2.47}$$

First an expression for $w(z)$ is found:

$$\int_Z^z \frac{\partial w}{\partial z}dz = \int_Z^z\left(-\frac{\partial u}{\partial x} - \frac{\partial v}{\partial y}\right)dz \tag{2.48}$$

$$w(z) - w(Z) = \int_Z^z\left(-\frac{\partial u}{\partial x} - \frac{\partial v}{\partial y}\right)dz \tag{2.49}$$

$$w(z) = \int_Z^z\left(-\frac{\partial u}{\partial x} - \frac{\partial v}{\partial y}\right)dz + w(Z) \tag{2.50}$$

$$w(z) = -\frac{\partial \int_Z^z u\,dz}{\partial x} - u(Z)\frac{\partial u}{\partial x} - \frac{\partial \int_Z^z v\,dz}{\partial y} - v(Z)\frac{\partial v}{\partial y} + w(Z) \tag{2.51}$$

$$w(z) = \underset{\textcircled{a}}{-\frac{\partial \int_Z^z u\,dz}{\partial x}} \underset{\textcircled{b}}{- \frac{\partial \int_Z^z v\,dz}{\partial y}} \tag{2.52}$$

$$\textcircled{a} = -\frac{\partial \int_Z^z \overline{u}r\left(\dfrac{z - Z}{H - Z}\right)dz}{\partial x} \tag{2.53}$$

$$\text{ⓐ} = -\frac{\partial (H - Z)\overline{u}\int_0^S r(s)\,ds}{\partial x} = -\frac{\partial (H - Z)\overline{u}R(S)}{\partial x} \tag{2.54}$$

$$\text{ⓐ} = -\frac{\partial (H - Z)\overline{u}}{\partial x}R(S) - (H - Z)\overline{u}\frac{\partial R(S)}{\partial x} \tag{2.55}$$

A similar expression can be found for ⓑ. The derivative of $R(S)$ with respect to x can be expanded as follows:

$$\begin{aligned}\frac{\partial R(S)}{\partial x} &= \frac{\partial R}{\partial S}\frac{\partial S}{\partial x}\\ &= r(S)\frac{1}{(H - Z)^2}\left(-\frac{\partial Z}{\partial x}(H - Z) - \frac{z - Z}{H - Z}\frac{\partial H}{\partial x} - \frac{\partial Z}{\partial x}\right)\end{aligned} \tag{2.56}$$

$$\frac{\partial R(S)}{\partial S} = \frac{r(S)}{H - Z}\left(-\frac{\partial Z}{\partial x} - S\frac{\partial (H - Z)}{\partial x}\right) \tag{2.57}$$

Applying Equation 2.57 to Equation 2.55 gives

$$\text{ⓐ} = \left(-\frac{\partial (H - Z)\overline{u}}{\partial x}\right)R(S) - (H - Z)\overline{u}\frac{r(s)}{(H - Z)}\left(-\frac{\partial Z}{\partial x} - S\frac{\partial (H - Z)}{\partial x}\right) \tag{2.58}$$

$$\text{ⓐ} = \left(-\frac{\partial (H - Z)}{\partial x}\overline{u} - \frac{\partial \overline{u}}{\partial x}(H - Z)\right)R(S) - r(S)\left(\overline{u}\frac{\partial Z}{\partial x} - \overline{u}S\frac{\partial (H - Z)}{\partial x}\right) \tag{2.59}$$

Solving similarly for ⓑ and substituting it in Equation 2.52, gives

$$w(z) = \nabla \cdot (h\mathbf{Q})R(S) - (\mathbf{Q} \cdot \nabla)Zr(S) + S(\mathbf{Q} \cdot \nabla)(H - Z) \tag{2.60}$$

Substituting Equation 2.60 in Equation 2.47 gives

$$\begin{aligned}\text{ⓒ} = \overline{u}\biggl(&\int_0^1 \nabla \cdot (h\mathbf{Q})R(S)r'(S)\,dS - \int_0^1 (\mathbf{Q} \cdot \nabla)Zr(S)r'(S)\,dS\\ &+ \int_0^1 S(\mathbf{Q} \cdot \nabla)hr'(S)\,dS\biggr)\end{aligned} \tag{2.61}$$

$$\text{ⓒ} = \overline{u}\biggl(\nabla \cdot (h\mathbf{Q})\int_0^1 R(S)r'(S)\,dS - (\mathbf{Q} \cdot \nabla)Z\int_0^1 r(S)r'(S)\,dS$$

$$+ (\mathbf{Q} \cdot \nabla)h \int_0^1 Sr'(S)\, dS \Big) \tag{2.62}$$

$$\text{Ⓒ} = \bar{u}(\nabla \cdot h\mathbf{Q})\gamma_1 - (\mathbf{Q} \cdot \nabla)Z\gamma_2 + (\mathbf{Q} \cdot \nabla)h\gamma_3) \tag{2.63}$$

where

$$\gamma_1 = r(1) - \int_0^1 r^2(s)\, ds$$

$$\gamma_2 = \frac{1}{2}(r^2(1) - r^2(0))$$

$$\gamma_3 = r(1) - 1$$

Integration of Ⓓ:

$$\text{Ⓓ} = \int_Z^H -\frac{1}{\rho}\frac{\partial p}{\partial x}\, dz \tag{2.64}$$

Under hydrostatic conditions, the term in the integral is constant, because

$$p(z) = \rho g(H - z) \tag{2.65}$$

and therefore

$$\frac{\partial p(z)}{\partial x} = \rho g \frac{\partial (H - z)}{\partial x} \tag{2.66}$$

Since the vertical coordinate z is independent of x and y,

$$\frac{\partial p(z)}{\partial x} = \rho g \frac{\partial H}{\partial x} \tag{2.67}$$

Then

$$\text{Ⓓ} = -g \int_Z^H \frac{\partial H}{\partial x}\, dz = -gh\frac{\partial H}{\partial x} \tag{2.68}$$

Integration of Ⓔ

$$\text{Ⓔ} = \frac{\mu}{\rho} \int_Z^H \underbrace{\frac{\partial^2 u}{\partial x^2}}_{\text{Ⓔ1}} + \underbrace{\frac{\partial^2 u}{\partial y^2}}_{\text{Ⓔ2}}\, dz \tag{2.69}$$

$$\text{(E1)} = \int_Z^H \frac{\partial^2 u}{\partial x^2}\,dz \tag{2.70}$$

Applying Leibnitz's rule twice, we write

$$\text{(E1)} = (H - Z)\frac{\partial^2 \bar{u}}{\partial x^2} + 2\frac{\partial \bar{u}}{\partial x}\frac{\partial (H - Z)}{\partial x} + \bar{u}\frac{\partial^2 (H - Z)}{\partial x^2}$$
$$-2\frac{\partial u(H)}{\partial x}\frac{\partial H}{\partial x} - u(H)\frac{\partial^2 H}{\partial x^2} + 2\frac{\partial u(Z)}{\partial x}\frac{\partial Z}{\partial x} + u(Z)\frac{\partial^2 Z}{\partial x^2} \tag{2.71}$$

$$\text{(E1)} = (H - Z)\frac{\partial^2 \bar{u}}{\partial x^2} + \left(2\frac{\partial \bar{u}}{\partial x}\frac{\partial (H - Z)\varepsilon}{\partial x} + \bar{u}\frac{\partial^2 (H - Z)\varepsilon}{\partial x^2}\right) \tag{2.72}$$

where

$$\varepsilon = 1 - \frac{r(1)H - r(0)Z}{H - Z}$$

Proceeding similarly with (E2), we obtain

$$\text{(E)} = \frac{\mu}{\rho}(h\nabla^2\bar{u} + 2(\nabla(h\varepsilon)\cdot\nabla)\bar{u} + Q\cdot\nabla^2(h\varepsilon)) \tag{2.73}$$

Integration of (F):

$$\text{(F)} = \frac{\mu}{\rho}\int_Z^H \frac{\partial^2 u}{\partial z^2}\,dz = \frac{\mu}{\rho}\frac{\bar{u}}{H - Z}\int_0^1 \frac{d^2 r}{ds^2}\,ds \tag{2.74}$$

$$\text{(F)} = \frac{\mu}{\rho}\frac{\bar{u}}{H - Z}(r'(1) - r'(0)) \tag{2.75}$$

$$\text{(F)} = \frac{\mu}{\rho}\frac{\bar{u}}{h}\phi \tag{2.76}$$

where $\phi = (r'(1) - r'(0))$.

The terms of Equation 2.30 can be integrated similarly. The integrated forms of Equations 2.29 and 2.30 then can be combined into a single expression:

$$h\frac{\partial Q}{\partial t} + \frac{\partial H}{\partial t}Q\alpha + h(Q\cdot\nabla)Q\beta - \frac{\partial H}{\partial t}Q\gamma_1 - Q(Q\cdot\nabla)Z\gamma_2 + Q(Q\cdot\nabla)h\gamma_3$$
$$= -hg\nabla H + h\frac{\mu}{\rho}\nabla^2 Q + \frac{\mu}{\rho}(2(\nabla(h\varepsilon))\cdot\nabla)Q + Q\nabla^2(h\varepsilon) + \frac{\mu}{\rho}\frac{Q}{h}\phi \tag{2.77}$$

After applying Equation 2.25 and dividing by h, we obtain

$$\frac{\partial Q}{\partial t} + \frac{\partial H}{\partial t}\frac{Q}{h}\alpha + (Q \cdot \nabla)Q\beta - \frac{\partial H}{\partial t}Q\gamma^1 - \frac{Q}{h}(Q \cdot \nabla)Z\gamma_2 + \frac{Q}{h}(Q \cdot \nabla)h\gamma_3$$
$$= -g\nabla H + \frac{\mu}{\rho}\nabla^2 Q + \frac{\mu}{\rho}\frac{1}{h}((2(\nabla(h\varepsilon)) \cdot \nabla)Q + Q\nabla^2(h\varepsilon)) + \frac{\mu}{\rho}\frac{Q}{h^2}\phi \tag{2.78}$$

where

$$\alpha = 1 - r(1)$$
$$\beta = \int_0^1 r^2(s)\,ds$$
$$\gamma_1 = r(1) - \beta$$
$$\gamma_2 = \tfrac{1}{2}(r^2(1) - r^2(0))$$
$$\gamma_3 = r(1) - 1$$
$$\varepsilon = 1 - \frac{1}{H - Z}(r(1)H - r(0)Z)$$
$$\phi = r'(1) - r'(0)$$

Equation 2.78 is the vertically integrated form of the two-dimensional flow model.

SELECTING A VELOCITY PROFILE

We now need to select a velocity profile. Coefficients α through ϕ in Equation 2.78 depend on the shape of the velocity profile, defined by the function r in Equations 2.31 and 2.32. The velocity profile r can be arbitrarily chosen, but additional assumptions about r can further simplify Equation 2.78. Figure 2.3 illustrates some possibilities, and Table 2.1 lists the corresponding coefficients for Equation 2.78.

In the simplest velocity profile, r is constant with depth (Fig. 2.3*a*). SEDSIM's versions use a constant-velocity profile, but several other alternatives are described here that could be implemented readily, proving enhanced capabilities at the expense of greater computing effort. One alternative (Fig. 2.3*b*) involves a stream profile in which the velocity is zero at the bottom, but immediately above and for the rest of the profile upward is constant. Another profile (Fig. 2.3*c*) involves a velocity of zero at both the bottom and top of the flow, with constant velocity elsewhere. Such a profile might be called a "turbidite" profile because it crudely approximates the profile of a submarine turbidity current moving through otherwise still water, with drag at both the bottom and top of the turbid flow. Such a profile is severely simplified because

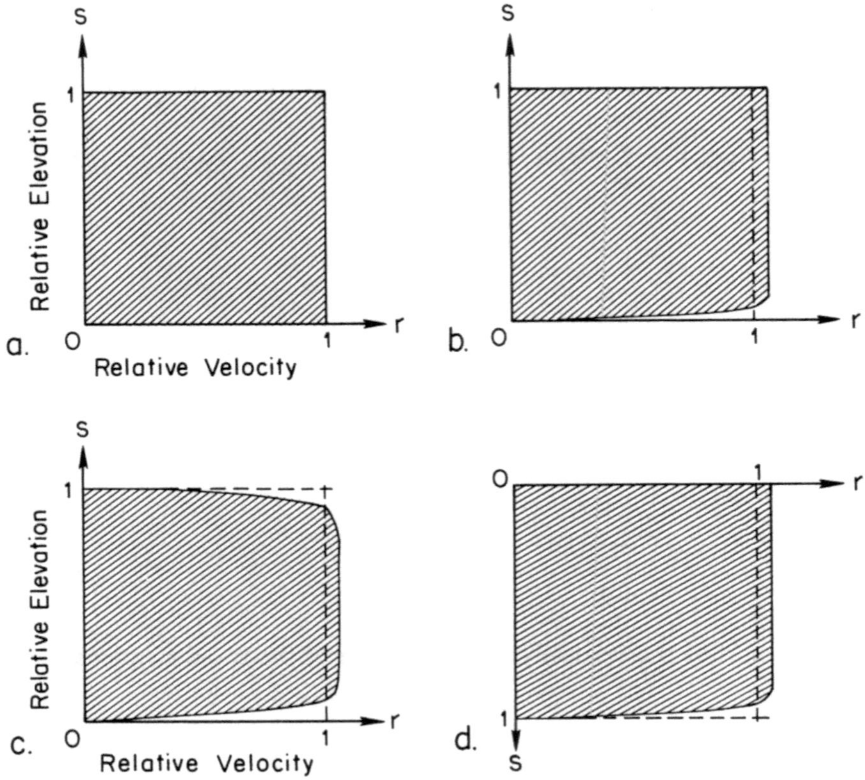

Figure 2.3 Example velocity profiles. (*a*) "Constant" profile incorporated in SEDSIM. (*b*) "Stream" profile. (*c*) "Turbidity" profile. (*d*) "Hypopycnal flow" profile.

TABLE 2.1 Coefficients of Two-Dimensional Kinematic Equation (Eq. 2.78) for Four Idealized Velocity Profiles

Coefficient	*1 Constant Profile*	*2 Stream Profile*	*3 Turbidity Current Profile*	*4 Hypopycnal Flow Profile*
α	0.	0.	1.	1.
β	~1.	~1.	~1.	~1.
γ_1	~0.	~0.	~1.	~1.
γ_2	0.	−0.5	0.	0.5
γ_3	0.	0.	1.	1.
ε	0.	$-Z/(H-Z)$	1.	$H/(H-Z)$
ϕ	0.	$-k$	~0.	k

it ignores vertical motion within the flow and displacement of and mixing with the surrounding fluid. Nevertheless, the profile is a useful approximation that yields results that appear realistic.

Finally, we can conceive of a velocity profile in which the velocity is zero at the free surface but is constant throughout the rest of the flow (Fig. 2.3*d*). We can invert this profile by assuming that the free surface is at the bottom of the flow and the rigid surface is at the top, as could be provided by a fluid of low density flowing over a stagnant fluid of slightly higher density. This profile can be called the "hypopycnal-flow" profile, following the nomenclature of Bates (1953) to describe rivers that discharge water of lower density into bodies of water of higher density, such as rivers flowing into the sea. For hypopycnal flow, topographic elevation Z must be replaced by sea level in the flow equations. Such a representation of hypopycnal flow is idealized because wind, waves, and mixing between the flow and the surrounding fluid are ignored.

When the flow is surrounded by or in contact with a medium of different density, as in turbidity currents and hypopycnal flow, Equation 2.78 must be modified by multiplying gravity by the relative density of the flow with respect to the surrounding medium (R_ρ), namely

$$R_\rho = \left| \frac{\rho_1 - \rho_2}{\rho_2} \right| \tag{2.79}$$

where

ρ_1 = density of the flow (river water)
ρ_2 = density of the surrounding fluid (seawater).

The profiles for streams, turbidity currents, and hypopycnal flow at river mouths, shown in Figures 2.3*b–d*, respectively, simplify the equations, but the computations are still lengthy for complicated flow situations. SEDSIM uses the constant profile (Fig. 2.3*a*) for all three flow situations because of its simplicity. For open channels, SEDSIM incorporates the constant-profile model directly. For turbidity flow and hypopycnal flow, however, Equation 2.79 is applied to correct the flow density. In addition, for hypopycnal flow SEDSIM employs sea level instead of topographic elevation in the continuity and momentum equations.

From the standpoint of fluid mechanics, the constant-profile assumption is too simple for some applications, but it is useful for many sedimentation simulation models. Other formulations, such as Bonham-Carter and Sutherland's (1968) "jet-flow" delta model, also assume a constant-velocity profile.

BOTTOM FRICTION

The constant-profile assumption implies there is no shear in any horizontal plane in the flow, and thus horizontal friction is absent within the flow. Friction must be reintroduced if the flow's behavior is to be realistic. Thus, we assume friction at the lower boundary for flow in streams, friction at both the lower and upper boundaries for turbidity currents, and friction at the lower boundary for hypopycnal flow.

It is convenient to employ empirical or semiempirical formulas to represent friction. In an open channel, bottom friction is proportional to the square of the average flow velocity. Therefore, acceleration due to bottom friction is given by

$$\mathbf{a} = -c_1 \frac{\mathbf{Q}|\mathbf{Q}|}{h} \tag{2.80}$$

where c_1 = bottom friction coefficient.

If flow is represented in only two horizontal dimensions, it is incorrect to use water viscosity μ in Equation 2.78, because the model applies only to turbulent flow at a macroscopic level, whereas μ applies to laminar flow or to microscopic fluid elements in turbulent flow. Therefore it is appropriate to replace μ by a coefficient c_2 that represents the effect of shear between portions of fluid of different horizontal velocities. If we assume a constant-velocity profile and add bottom friction, Equation 2.78 becomes

$$\frac{\partial \mathbf{Q}}{\partial t} + (\mathbf{Q} \cdot \nabla)\mathbf{Q} = -g\nabla H + \frac{c_2}{\rho}\nabla^2\mathbf{Q} - c_1\frac{\mathbf{Q}|\mathbf{Q}|}{h} \tag{2.81}$$

In Lagrangian form,

$$\frac{D\mathbf{Q}}{Dt} = -g\nabla H + \frac{c_2}{\rho}\nabla^2\mathbf{Q} - c_1\frac{\mathbf{Q}|\mathbf{Q}|}{h} \tag{2.82}$$

SPECIFYING BOUNDARY CONDITIONS FOR FLOW

The equations of flow can be utilized, provided that we specify the boundary conditions, including the initial conditions when the simulation begins. The initial conditions must provide the flow velocity and depth at every point at time 0:

$$\mathbf{Q}(x, y, t_0) = \mathbf{Q}_0(x, y) \tag{2.83}$$

$$h(x, y, t_0) = h_0(x, y) \tag{2.84}$$

The boundary conditions can include the "shoreline" (the line where flow depth h becomes zero) as well as the edges of the area. At the shoreline,

$$h(x, y, t) = 0 \tag{2.85}$$

and therefore

$$\bar{u}_s \frac{\partial h_s}{\partial x} + \bar{v}_s \frac{\partial h_s}{\partial y} + \frac{\partial h_s}{\partial t} = \frac{Dh_s}{Dt} = 0 \tag{2.86}$$

where

$\bar{u}_s, \bar{v}_s$ = components of velocity at the shoreline
h_s = flow depth at the shoreline
D = derivative following motion of the fluid

At the edges of the simulated area, all fluid is removed, and the flow depth becomes zero, which approximates the physical condition that permits the flow to exit freely at boundaries:

$$h_b = 0 \tag{2.87}$$

The boundary conditions also include fluid that is supplied at "sources" whose geographic locations are specified. Each source occupies a small area and the flow rate at each point within a source is defined by

$$\nabla \cdot (h(x_{s_i}, y_{s_i})Q(x_{s_i}, y_{s_i})) + \frac{\partial h}{\partial t} = F_{s_i} \tag{2.88}$$

where

x_{s_i}, y_{s_i} = coordinates defining geographic positions of points within source
s_i = source number
F_{s_i} = flow rate per unit area at source s_i (integration over the area of the source yields total fluid input)

The velocity at each source is provided by

$$Q(x_{s_i}, y_{s_i}) = Q_{s_i} \tag{2.89}$$

where Q_{s_i} = velocity at source s_i.

With initial and boundary conditions defined by Equations 2.83–2.89, the flow system is completely defined. The remaining challenge is to obtain the solution that will yield depths of flow and horizontal flow velocities as functions of position and time. The existence and uniqueness of a solution to the system of equations are not rigorously treated here, but we illustrate the solution with examples.

SUMMARY

Equation 2.82 and the continuity equation (Eq. 2.25) constitute the systems of equations on which flow is based. Coefficients c_1 and c_2 need to be calibrated, as explained later. Equation 2.82 defines acceleration of each element of fluid, the terms on the right side corresponding, respectively, to acceleration caused by slope of the water surface, "lateral" friction within the fluid, and friction against the lower or upper boundary of the flow.

Equations 2.25 and 2.82 have been derived from the Navier-Stokes equations, making the following simplifying assumptions: (1) Flow is always confined between a rigid impermeable surface and a free surface, and neither surface can have vertical or overhanging slopes. (2) Flow velocity is constant along any vertical line through the flow. (3) Pressure distribution is hydrostatic. (4) Friction at the flow boundaries is proportional to flow velocity squared.

CHAPTER 3

Numerical Solutions of the Flow Equations

The flow equations are written initially in differential form, and the task before us is to solve them. Analytical solutions are difficult, but numerical methods can provide approximate solutions that are directly usable. In numerical solutions, continuous variables are represented only at discrete points. For example, a continuous function defined on a plane can be replaced by a function represented by a finite set of points arranged on a grid in the plane. The differential equations representing the continuous function then can be replaced by algebraic expressions that represent the variables in discrete form. The succession of evaluations of the discrete variables using the algebraic expressions at grid points leads to the numerical solution, consisting of the numerical values arranged on the grid. If time is involved, it can be divided into discrete intervals or steps. The numerical solution then consists of a set of grids, one for each time increment.

CONSISTENCY, ACCURACY, STABILITY, AND EFFICIENCY

In numerical solutions, we must generally choose from among various algebraic approximations. Our selection should be influenced by criteria that include consistency, accuracy, stability, and efficiency. *Consistency* in a numerical approximation pertains to whether or not a numerical approximation approaches the exact analytical solution of the differential equations as the discrete intervals (the time steps and the grid elements) are made smaller and smaller. *Accuracy* is a measure of closeness between a value given by an approximate algebraic expression at a specific point and the value provided by the exact solution at that point. *Stability* is a measure of the degree to which errors progressively decrease during a succession of iterations. *Efficiency* pertains to the ability of the numerical method to reach a suitable approximation in an acceptable number of iterations. Consistency and stability are the most important properties, and when both are satisfactory, the numerical method is said to be *convergent*.

Storage of information is an important consideration in obtaining a solution. It pertains to the number of variables that must be retained when arithmetic operations are performed. There is a trade-off between storage and efficiency. Some numerical methods can be made more efficient at the expense of requiring that more information be stored. Storage affects the amount of the computer's memory that must be dedicated to the solution, whereas efficiency affects the amount of computer time or number of arithmetic cycles that are needed to obtain a solution. Thus we must choose among conflicting alternatives in the quest for the most suitable procedure.

STATIC VERSUS DYNAMIC MODELS

Numerical solutions for static systems are simpler than for dynamic systems. For a static system the numerical solution is obtained by progressively iterating through the equations representing the system until values at grid points approximate the solution. The final values in the gridwork represent the solution (Table 3.1); and previous values can be discarded. Such a procedure is not feasible for a dynamic system because there is no "final" set of values. We can solve the dynamic system by starting with grid values representing the initial state of the system, and then iterating forward to define the subsequent states that correspond with a succession of time increments. Each iteration then can represent a single time increment, and the succession of values obtained is the solution. The solution is potentially infinite in length, because time never ends and dynamic systems constantly change. Solutions to dynamic systems require greater accuracy and efficiency at each iteration, because every iteration, not just the "last" one, is part of the numerical solution. Thus, dynamic systems are more computationally challenging than static systems of equivalent detail.

INTERPOLATING GRID VARIABLES

SEDSIM's flow equations are solved by a particle-mesh method that employs grids to represent variables at fixed points. The grids represent topographic elevation, flow depth, and flow velocity. Particles of fluid or fluid

TABLE 3.1 Comparison of Numerical Solutions Used with Static and Dynamic Spatial Models That Employ Successive Iterations

Static Model

Iteration 0		Iteration 1		Iteration 2	···	Iteration n
Grid 0	→	Grid 1	→	Grid 2	→	Grid n
						(solution)

Dynamic Model

Iteration 0	Grid 0 (t = 0)	
	↓	
Iteration 1	Grid 1 (t = 1)	
	↓	(solution)
Iteration 2	Grid 2 (t = 2)	
···	↓	
Iteration m	Grid m (t = m)	

elements are represented as variables at points whose positions move with the fluid. Thus, flow velocity is represented both as a grid variable and as a fluid-element variable. There are various ways of representing two-dimensional variables with regular grids. The cells of the grid may be square, rectangular, triangular, or hexagonal. SEDSIM uses square cells.

Functions represented by grids in SEDSIM are continuous. Topographic surfaces and depth of flow are assumed to be continuous. The representation of these continuous functions at every point where fluid elements may be located requires interpolation between grid points. The interpolation function can be represented graphically by a surface whose elevation varies within each grid cell. Figure 3.1 shows several alternatives for interpolation. If square or rectangular cells are employed, curved surfaces for interpolation are needed if discontinuities are to be avoided. Curved surfaces, however, require more computation than planes, which are simple to compute but which yield discontinuities between cells. A compromise is needed.

First, consider the use of planes for interpolation. A plane represents a linear interpolation, and the simplest arrangement involves cells that are square and have horizontal surfaces (Fig. 3.1*a*), so the interpolation value is a constant for each cell. Discontinuities are severe, however. A compromise is to use inclined planes (Fig. 3.1*b*) so that interpolation is linear and computationally simple, and the severity of the discontinuities is reduced. Triangular cells permit planes to join without discontinuities (Fig. 3.1*c*), but, among other problems, triangular cells require trigonometric functions for determining locations of points within cells, thus increasing computing effort. There is no

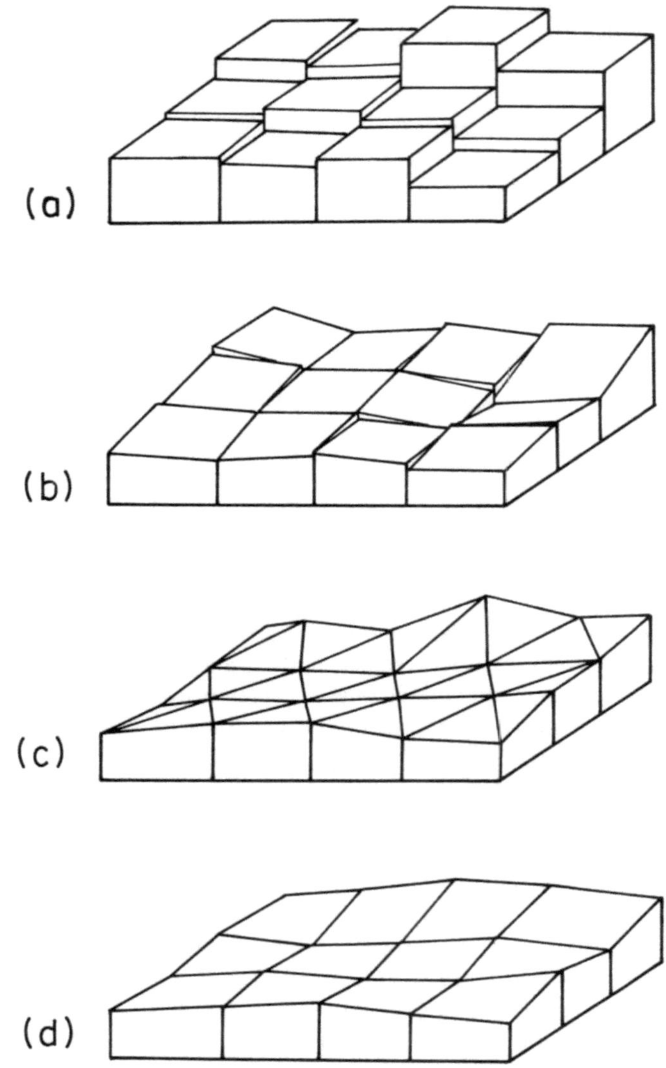

Figure 3.1 Example grid interpolation functions. (*a*) Square cells with constant interpolation function. (*b*) Square cells with linear interpolation function. (*c*) Triangular cells with linear interpolation function. (*d*) Square cells with hyperbolic-paraboloid interpolation function (used by SEDSIM). Hyperbolic paraboloid fits cell corners exactly and yields straight lines at boundaries shared by adjacent cells, thus avoiding discontinuities between cells.

wholly satisfactory procedure, but the best compromise involves square cells and a curving interpolation function consisting of hyperbolic paraboloids (Fig. 3.1*d*) that fit each cell's vertices exactly, avoid discontinuities at cell edges, and permit the value of the function to be calculated within a cell as follows:

$$z(Z_1, Z_2, Z_3, Z_4, X, Y) = Z_1XY + Z_2(1 - X)Y + Z_3X(1 - Y) + Z_4(1 - X)(1 - Y) \quad (3.1)$$

where

z = variable being evaluated
Z_i = value of z at each corner of a cell
X, Y = coordinates of position within cell ($0 \leqslant X \leqslant 1, 0 \leqslant Y \leqslant 1$)

The slope of the function represented by the grid can be expressed with respect to the x and y directions (such as east-west and north-south) by employing first derivatives at a point (x, y), one parallel to the x axis and the other to the y axis:

$$S_x(Z_1, Z_2, Z_3, Z_4, X, Y) = (Z_2 - Z_1)Y + (Z_3 - Z_4)(1 - Y)$$
$$S_y(Z_1, Z_2, Z_3, Z_4, X, Y) = (Z_3 - Z_1)X + (Z_4 - Z_2)(1 - X) \quad (3.2)$$

where S_x, S_y = slopes of z in the x and y directions.

FLOW DEPTH

Flow depth is proportional to the number of fluid elements present in each cell (Fig. 3.2). Each fluid element has a fixed volume, so the average flow depth over an area is calculated from the number of fluid elements within the area, the unit volume of an element, and the size of the area. Such a relationship is intuitively obvious, and it also can be deduced from the continuity equation (Eq. 2.25). In fact, by integrating Equation 2.25 over the area of a cell and then applying Stokes' theorem, we obtain

$$\frac{\partial h}{\partial t} = \frac{1}{A} \oint_{\partial A} h(\mathbf{N} \cdot \mathbf{Q})\, dl \quad (3.3)$$

where

A = area of one cell
$\oint$ = curvilinear integral
∂A = perimeter of one cell
$\mathbf{N}$ = vector of unit length extending perpendicular and outward to ∂A
l = length

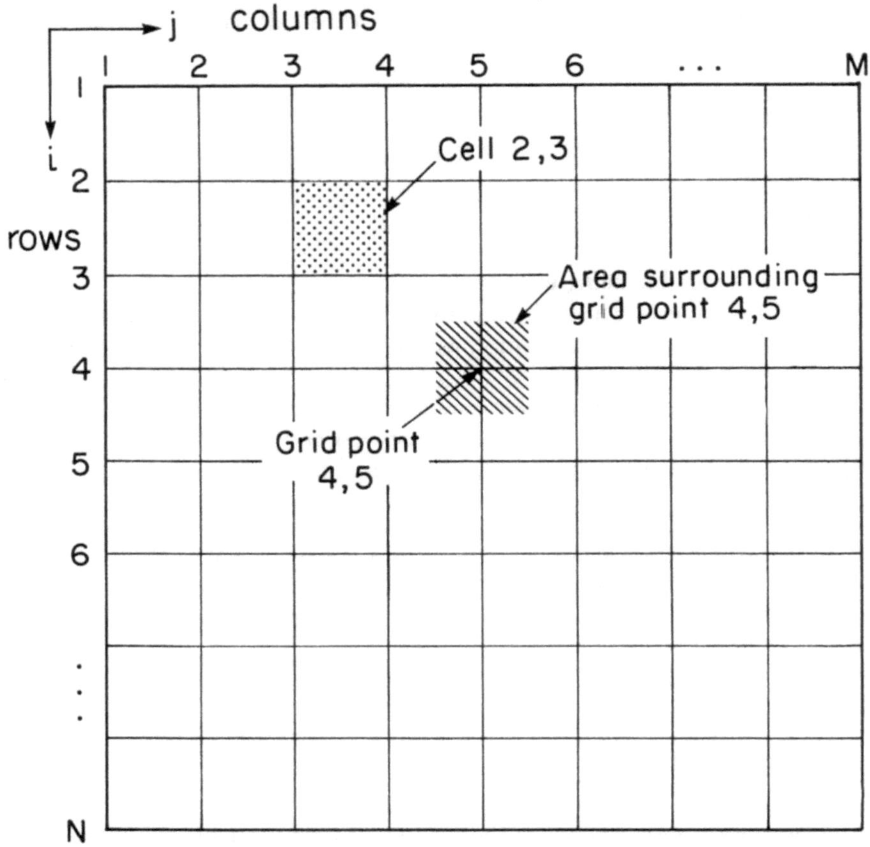

Figure 3.2 Two-dimensional grid design employed by SEDSIM. Flow parameters at grid points represent averages within square area surrounding each grid point.

Then, integration of Equation 3.3 over time t yields

$$h_t - h_0 = \frac{1}{A} \int_0^t \left(\oint h(\mathbf{N} \cdot \mathbf{Q})\, dl \right) dt \tag{3.4}$$

or

$$h_t = h_0 + \frac{V}{A} \tag{3.5}$$

where V = net volume of fluid that has entered the cell since time 0.

Equation 3.5 can be expressed in discrete algebraic form:

$$h_{i,j} \simeq \frac{V_e L_{i,j}}{A} \tag{3.6}$$

where

$h_{i,j}$ = flow depth at grid point i,j
V_e = volume of one fluid element
$L_{i,j}$ = number of fluid elements nearest grid point i,j (Fig. 3.2)
A = area

In SEDSIM, A is assumed to be the area consisting of a square with a grid point at its center (Fig. 3.2). Thus a grid representing depth of flow is provided by using Equation 3.6.

VELOCITY OF FLOW

Flow velocity at each grid point is calculated as the average velocity of the fluid elements in the vicinity of the grid point:

$$\mathbf{Q}_{i,j} = \frac{1}{L} \sum \mathbf{Q}_k \tag{3.7}$$

where

$\mathbf{Q}_{i,j}$ = velocity at grid point i,j
Σ = summation over k, $1 \leq k \leq L$
L = number of fluid elements near grid point i,j
$\mathbf{Q}_k$ = velocity of fluid element k

The value of $\nabla^2\mathbf{Q}$ can be calculated from the average flow velocity at grid point ($\mathbf{Q}_{i,j}$), and the average velocity in the surrounding grid points, as is standard in finite-difference schemes:

$$\nabla^2\mathbf{Q} \simeq \frac{(\mathbf{Q}_{i,j+1} + \mathbf{Q}_{i,j-1} + \mathbf{Q}_{i+1,j} + \mathbf{Q}_{i-1,j})/4 - \mathbf{Q}_{i,j}}{(\Delta X)^2} \tag{3.8}$$

FLUID-ELEMENT VARIABLES

Fluid position and velocity are represented in two horizontal dimensions as a set of moving fluid elements or moving points. The acceleration of each fluid element can be approximated as

$$\frac{D\mathbf{Q}}{Dt} \simeq \frac{\Delta \mathbf{Q}}{\Delta T} = \frac{\mathbf{Q}_{t+1} - \mathbf{Q}_t}{\Delta T} \tag{3.9}$$

Using Equation 3.9, we can write Equation 2.82 algebraically as

$$\frac{\mathbf{Q}_{k,t+1} - \mathbf{Q}_{k,t}}{\Delta T} = g\mathbf{S}_k - c_1 \frac{\mathbf{Q}_k|\mathbf{Q}_k|}{Z_k - H_k} + \frac{c_2}{\rho} \frac{(\mathbf{Q}_{i,j+1} + \mathbf{Q}_{i,j-1} + \mathbf{Q}_{i+1,j} + \mathbf{Q}_{i-1,j})/4 - \mathbf{Q}_{i,j}}{(\Delta \mathbf{X})^2} \tag{3.10}$$

where

H_k = water surface elevation at location of element k
$\mathbf{S}_k$ = water surface slope at location of element k
Z_k = topographic elevation at location of element k

Let $\mathbf{C}$ be the term on the right side of Equation 3.10. Then the velocity of each fluid element is

$$\mathbf{Q}_{k,t+1} = \mathbf{Q}_{k,t} + \mathbf{C}\,\Delta T \tag{3.11}$$

The position of each fluid element is

$$\mathbf{X}_{k,t+1} = \mathbf{X}_{k,t} + \frac{\mathbf{Q}_{k,t} + \mathbf{Q}_{k,t+1}}{2} \Delta T \tag{3.12}$$

where

$\mathbf{X}_{k,t}$ = position of fluid element k at time t
$\mathbf{X}_{k,t+1}$ = position of fluid element k at time $t + 1$

Equations 3.11 and 3.12 are used to update velocities and positions of each fluid element at each time increment.

SUMMARY

The differential equations describing flow that were derived in Chapter 2 have been expressed in discrete form, making use of two techniques: (1) Fluid position and velocity have been represented by a number of moving fluid elements. (2) Variables that can be evaluated at fixed points have been represented on a grid with square cells. These variables include flow velocity (which is also represented as a fluid-element variable), topographic elevation, flow depth, and the spatial derivatives of velocity, elevation, and depth. At each new time increment, the discrete equations provide expressions for (a) grid variables in terms of other grid variables (Eq. 3.8), (b) grid variables in terms of fluid-element variables (Eqs. 3.6 and 3.7). and (c) fluid-element variables in terms of grid variables (Eqs. 3.11 and 3.12). In case (c), the grid variables must be evaluated at the precise location of each fluid element, thus requiring interpolation between grid points. The interpolation function consists of a hyperbolic paraboloid fitted to each cell (Eqs. 3.1 and 3.2). Successive equations of the discrete expressions then yield an approximation of the solution of the flow equations. The next chapter describes implementation of these procedures in computer program SEDCYC1.

CHAPTER 4

SEDCYC1: SEDSIM's Program for Flow

As described earlier, SEDSIM's computer programs are organized in three levels. Names of the programs employ suffixes 1, 2, or 3 to indicate the level. The first level (SEDSIM1) simulates flow only and uses various accessory programs also labeled with suffix 1. The second (SEDSIM2) deals with a single sediment type, and the third (SEDSIM3) deals with up to four types of sediment. Each of the three groups contains several main programs and their subroutines.

The most important program within SEDSIM1 is SEDCYC1. Program SEDCYC1 simulates flow in conjunction with various accessory programs that prepare data for input to SEDCYC1 and also handle graphic and printed output from SEDCYC1. Figure 4.1 shows relationships between the programs.

SEDCYC1 requires an input data file that provides information about the initial state of the flow system to be simulated. Table 4.1 is an example file and contains physical parameters, a topographic grid, and specifications for sources of fluid. Input files are prepared with program SEDINI1, which interactively generates grids and writes pertinent information in appropriate formats.

SIMULATION CYCLES WITH SEDCYC1

SEDCYC1 can be executed after an input data file has been prepared. SEDCYC1 first calls subroutine READDF1, which reads the input data file and then enters the main loop. Each cycle in the loop represents a single time in-

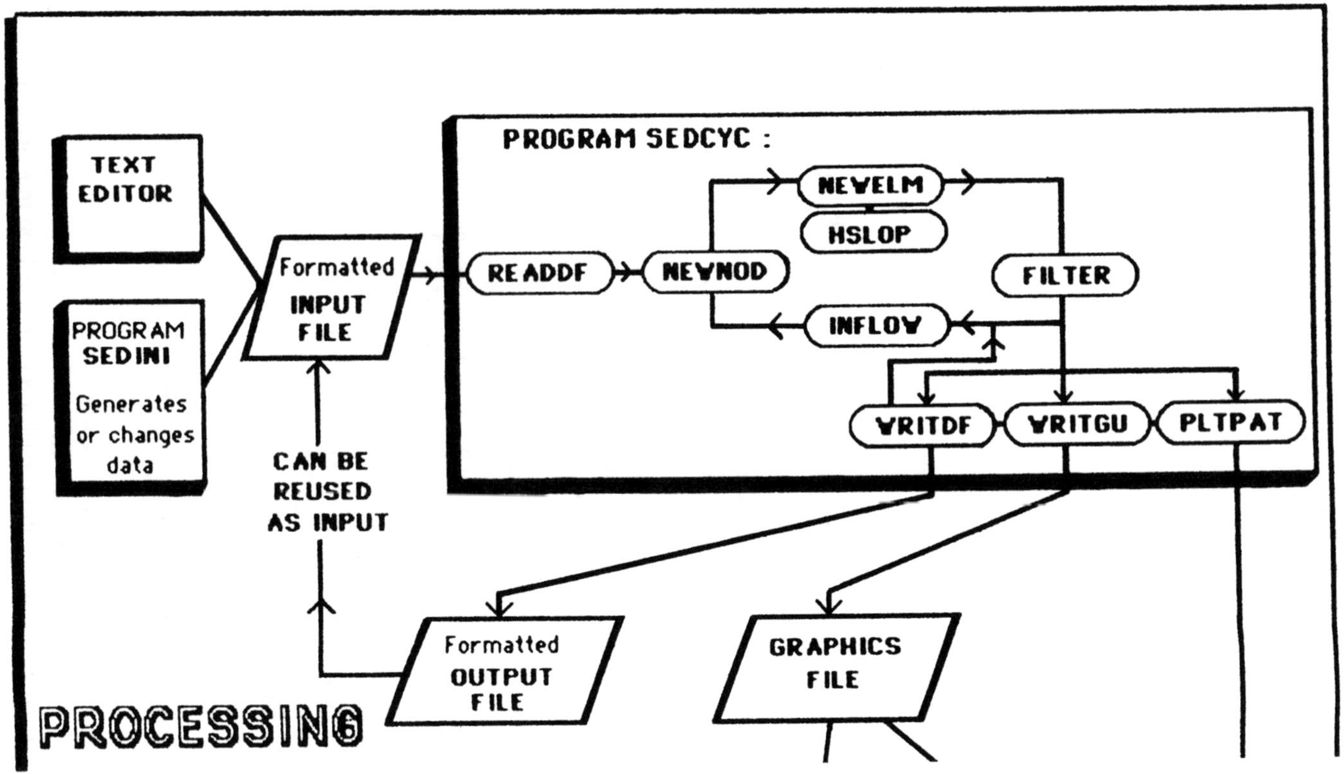

TEXT EDITOR
PROGRAM SEDINI
Generates or changes data
Formatted INPUT FILE
CAN BE REUSED AS INPUT
PROGRAM SEDCYC :
READDF
NEWNOD
NEWELM
HSLOP
INFLOW
FILTER
WRITDF
WRITGU
PLTPAT
Formatted OUTPUT FILE
GRAPHICS FILE
PROCESSING

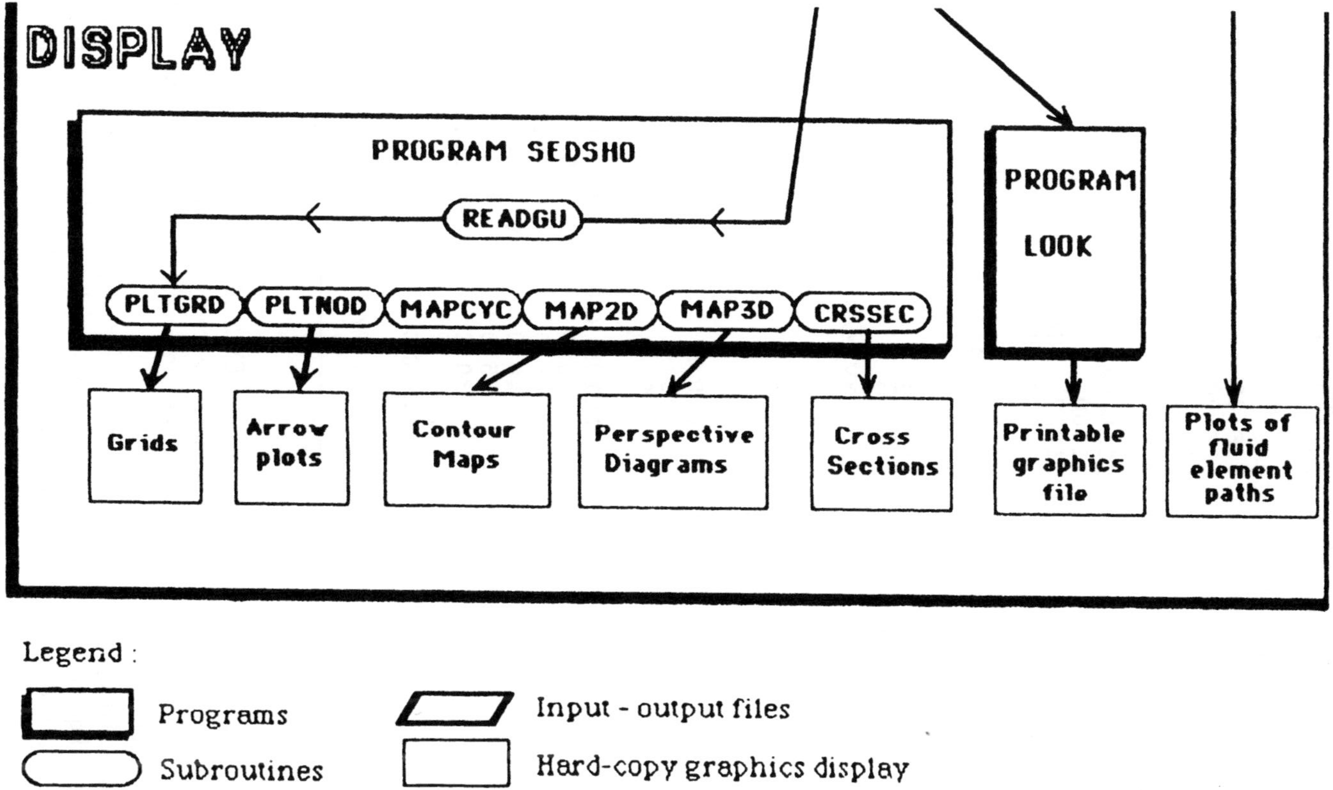

Figure 4.1 Generalized flowchart of program elements used in SEDSIM's three versions.

TABLE 4.1 Example Input Data File for SEDCYC1 in Experiment Involving Flow over Inclined Plane

```
TITLE: FLOW ON SLOPING SURFACE

RUN PARAMETERS:
START TIM=0.00E+00 Y
END TIM   =0.01E+00 Y
DISPL INT=0.01E-01 Y
TIM INCR1=0.50E+01 S

GENERAL PHYSICAL PARAMETERS:
FLOW DENS=    1000. KG/M3
SEA DENS =    1027. KG/M3
LAT FRICT=    100.0 NS/M2
ROUGHNESS=    0.040
EL VOLUME=0.01E+06 M3

SOURCES:
INTERVAL =0.05E-03 Y
N SOURCES=        3
   X(M)      Y(M)     XV(M/S)  YV(M/S)
     750.    2675.        0.5
    1000.    2675.       -0.5
     875.    2675.        0.

TOPOGRAPHY:
GRID SIDE=    100.0 M
NROWS     =       5
NCOLS     =       5

GRID NODES ELEVATION (SURFACE) (M)
    50.0      50.0      50.0      50.0      50.0
    45.0      45.0      45.0      45.0      45.0
    40.0      40.0      40.0      40.0      40.0
    35.0      35.0      35.0      35.0      35.0
    30.0      30.0      30.0      30.0      30.0

ELEMENT POSITIONS AND VELOCITIES:
N OF ELEM=        0
   X(M)      Y(M)     XV(M/S)  YV(M/S)
```

crement. During each cycle, subroutine NEWNOD1 updates the grid variables, and subroutine NEWELM1 updates the particle variables, NEWNOD1 and NEWELM1 being alternately executed in leapfrog fashion (Hockney and Eastwood, 1981).

NEWNOD1 updates the flow velocity and flow depth at each grid point according to Equations 3.6 and 3.7. It calculates depth from the number of fluid elements in the square area around each grid point (Fig. 3.2). Subroutine

NEWNOD1 assigns each fluid element to the nearest grid point and then counts the number of elements at that grid point. The number of fluid elements at each grid point, multiplied by the volume of a single element, and divided by the area surrounding each grid point (which is the area of a cell) gives the flow depth at that grid point (Eq. 3.6). Similarly, velocity at each grid point is found by averaging the velocities of the elements assigned to that grid point.

Subroutine NEWELM1 updates the fluid-element variables. Each fluid-element variable is determined by its previous value and the present values of the grid variables at the point where the fluid element is located (Eqs. 3.11 and 3.12). Because the elevation, flow depth, and velocity are required at the location of each particle that is to be updated, interpolation from the cell's four corners is necessary and is carried out by subroutine HSLOP1, employing Equations 3.1 and 3.2.

The boundary conditions at fluid sources are specified by subroutine INFLOW1, which provides new fluid elements according to information contained in the input data file. Subroutine FILTER1 applies a "time filter" to the grid variables. This time filter averages the values of a variable over a given interval of time, thus eliminating erratic small variations in fluid-element velocities prior to graphic display.

At preselected times during the execution of the main loop, subroutine WRITGU1 writes a graphics file containing a summary of the state of the system. The simulated time between successive calls of WRITGU1 is controlled by the "display interval" specified in the input data file. The graphics file provides a record of the intermediate states during a simulation experiment, which may be displayed with program SEDSHO1. When a simulation experiment is complete, subroutine WRITDF1 writes the final state of the system in the same format as the original input so that the results can serve as input for succeeding experiments, if desired.

GRAPHIC OUTPUT

Defining the state of the flow model at any specific moment requires that much information be stored, including topography, flow velocities, flow depths, and positions of sources of fluid. This information is displayed graphically by SEDSHO1, which reads the graphics file written by SEDCYC1 and converts it for display on a color graphics terminal or color plotter. The graphics file can grow large if the simulation run is long and information is requested at short intervals. Therefore the graphics file cannot contain all information necessary to describe each state of the system at each time. For economy of storage, the graphics file does not include information on individual fluid elements. Instead the graphics file contains two grids, one representing flow velocities and the other recording flow depths.

SEDSHO1 interactively asks the user what displays are desired and what span of time they are to represent. The displays include topographic contour maps or "fishnet" diagrams, current-velocity plots, and depth plots. It is useful to display a quick succession of flow-velocity plots at brief intervals on the terminal to document interactions of unsteady flows, much as a movie or video display would show them.

If a graphics terminal is not available, program LOOK1 will convert graphics files into formatted numerical displays for printing on a line printer, where printouts of flow velocities and depths are presented as arrays.

It may be useful to display information while a simulation experiment is in progress rather than await its completion. For example, we may wish to view the paths of the fluid elements in experiments that involve only few fluid elements. Subroutine PLTPAT1 plots arrows or vectors showing the displacement of each fluid element during each time increment. Examples in Figures 4.2, 4.3, and 4.4 were generated with PLTPAT1. When linked with a color graphics terminal, PLTPAT1 shows the fluid elements' movements while the simulation program is running.

VERIFICATION THAT SEDCYC1 CONSERVES ENERGY AND MOMENTUM

We need assurance that SEDCYC1 accords fully with the dictum that energy and momentum be conserved. We tested SEDCYC1 at varying levels of complexity to ensure that the test criteria were met when SEDCYC1 was operated under conditions known to conserve energy and momentum. The first series of tests verified that SEDCYC1's method for representing flow does not introduce any spurious losses or gains in energy when friction is absent. Equations 2.25 and 2.82 are shown to conserve energy when friction coefficients c_1 and c_2 are zero.

In one experiment, a single fluid element with a mass of 100 kg was allowed to move freely under the influence of gravity for a simulated span of 500 s. The experiment involved flow over a surface shaped as an inverted square pyramid whose dimensions were 2000 m on each side and 30 m deep (Fig. 4.5). Time increments of 1 s were employed, and a vector was drawn after each increment by subroutine PLTPAT1, thereby plotting the path of the fluid element (Fig. 4.2). The length of each vector is proportional to the element's velocity during the time increment represented by the vector. As Figure 4.2 reveals, the fluid element moved back and forth over the surface, with neither loss nor gain of energy, documenting that there was no variation in total energy. Of course, potential energy and kinetic energy are interchanged repeatedly during the experiment, but there is no net change in the sum of the two forms. By contrast, when friction was incorporated, the amplitudes of the fluid element's

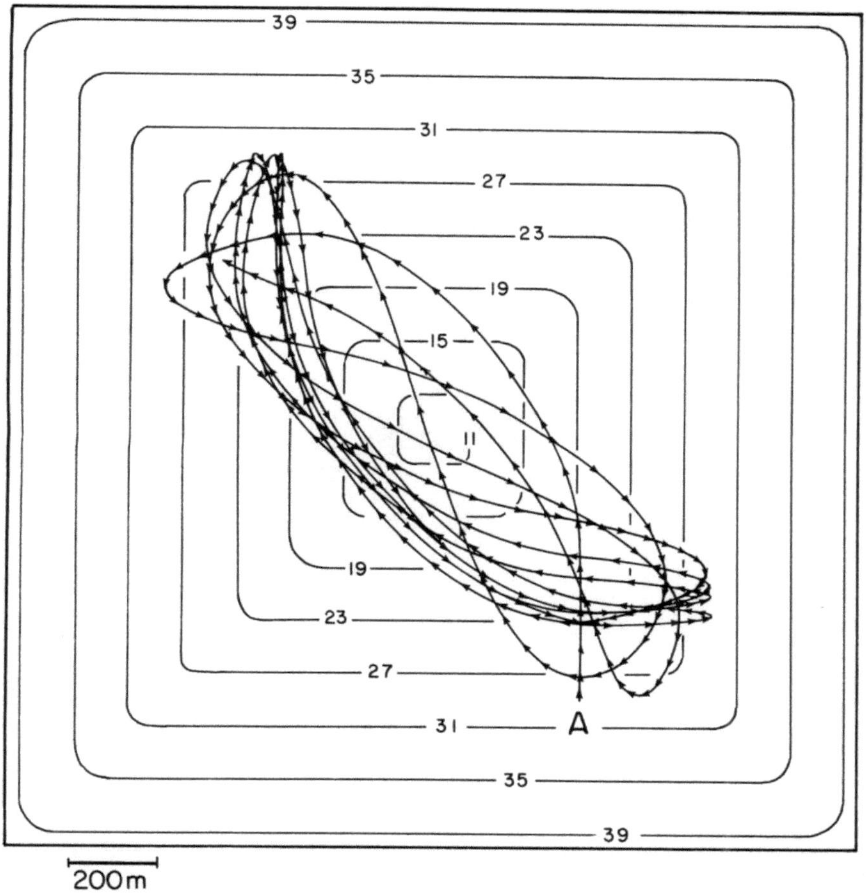

Figure 4.2 Contour map of basin in Figure 4.5 showing path of single fluid element after release from point A. Friction is absent and element moves back and forth endlessly. Velocity, indicated by arrow lengths at intervals of 5 s, varies such that sum of fluid element's kinetic energy plus its potential energy remains constant, documenting that energy is conserved. Contours are in meters.

movement progressively decreased and eventually died out (Fig. 4.3), representing progressive transformation of potential and kinetic energy to heat.

When more than one fluid element is present, both the elevation of the water's surface and the friction between elements as they move against each other affect motion. On a frictionless horizontal surface, Equations 2.25 and 2.82 conserve momentum. Although conservation of momentum is not strictly represented by the numerical method employed if only two fluid elements are present, momentum is conserved when many fluid elements are present.

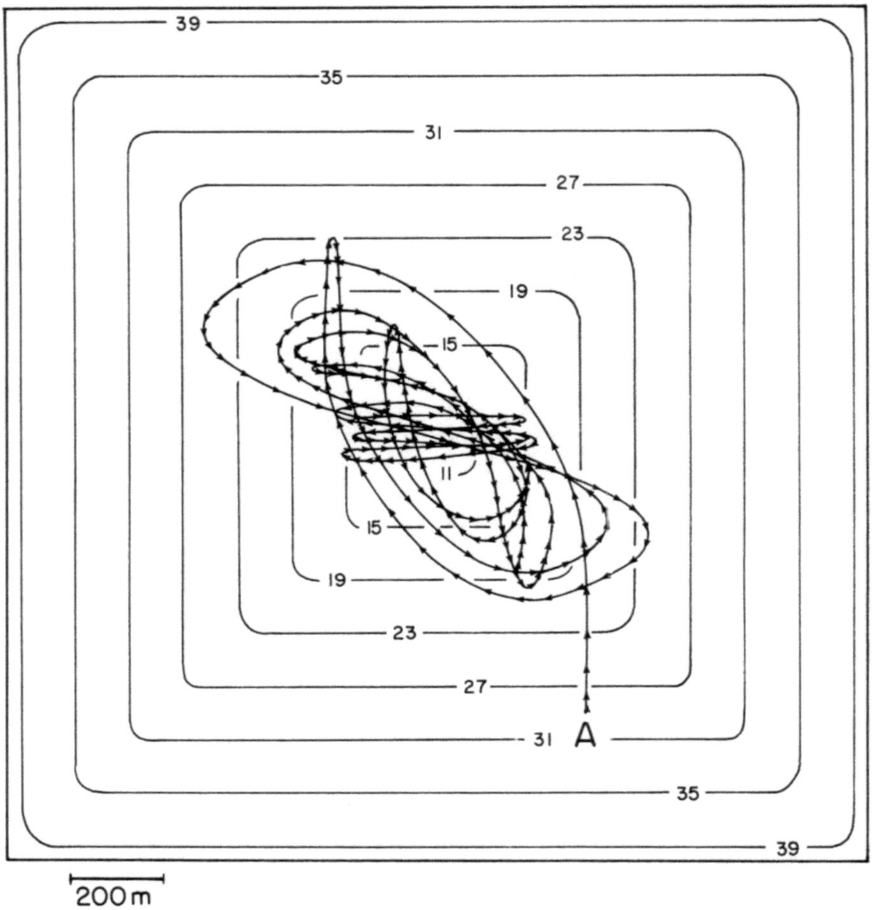

Figure 4.3 Path of single fluid element in experiment similar to that in Figure 4.2, except that fluid element is slowed by friction proportional to square of its velocity.

The test for conservation of momentum involved a second series of experiments in which two groups of fluid elements were present on an horizontal surface, with no bottom friction (Fig. 4.4). The two groups were released along paths that intersected each other. Momentum is conserved if the center of gravity of the overall system moves uniformly along a straight line. The paths of the centers of gravity of each of the two groups are shown with dashed heavy lines, whereas the path of the overall center of gravity is shown with a bold solid line. Figure 4.4 documents that the center of gravity moved in a uniform straight line.

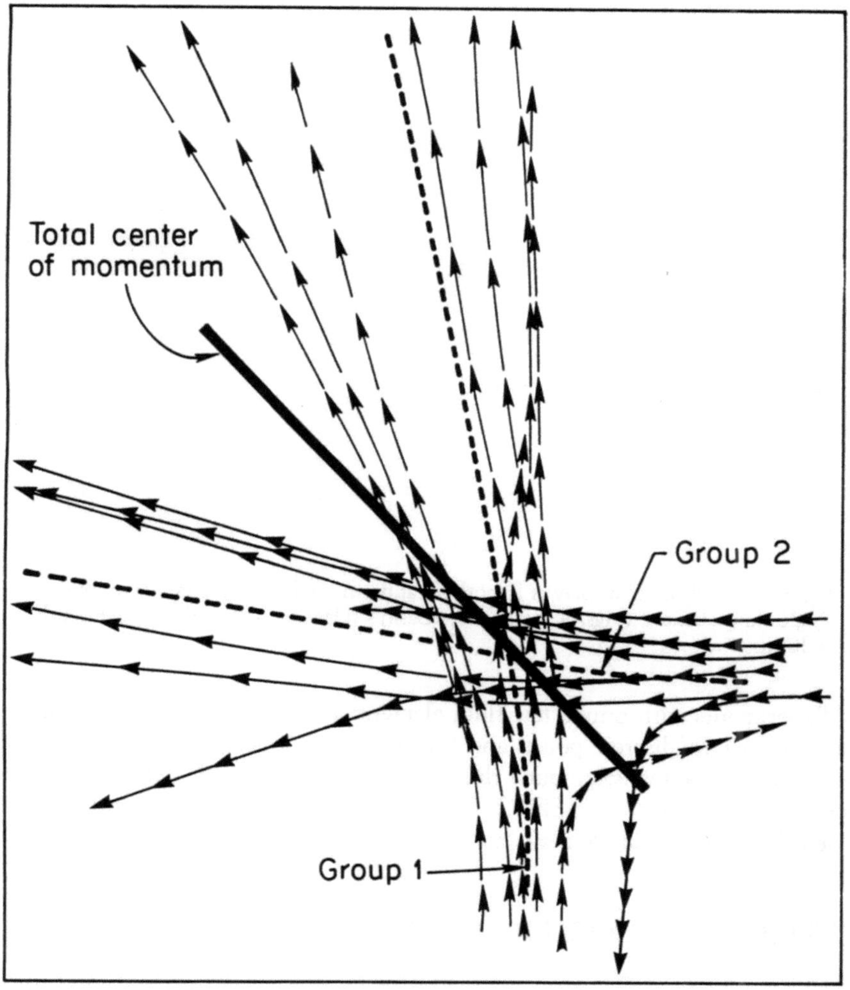

Figure 4.4 Map of trajectories of two groups of fluid elements moving on horizontal surface documenting that momentum is conserved. Two groups of fluid elements were released with initial trajectories that intersect each other. Arrows show paths of individual elements. Lengths between arrowheads are proportional to velocity. Paths of centers of gravity of each group (heavy dashed lines) deviate due to encounter between groups, but path of overall center of gravity (bold solid line) moves uniformly along straight path, demonstrating that momentum is conserved.

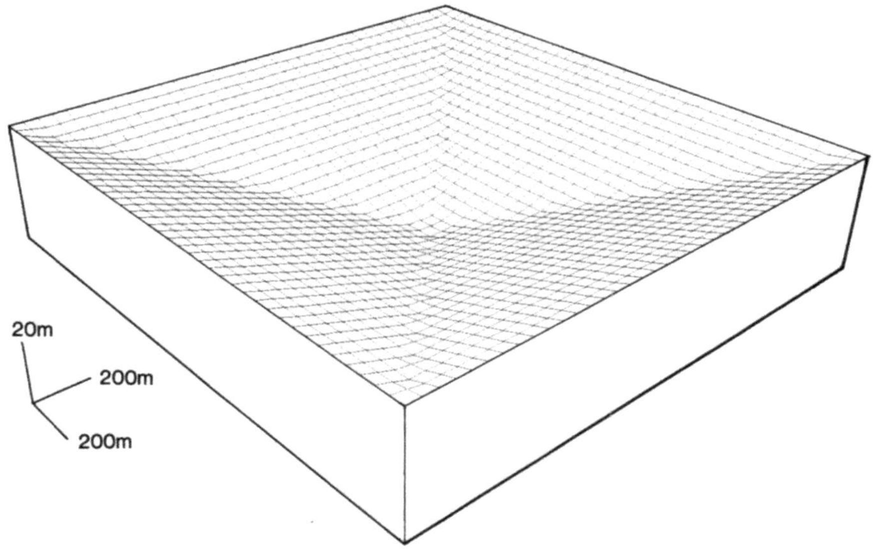

Figure 4.5 "Fishnet" display of hypothetical basin used for experiments involving single fluid element (Figs. 4.2 and 4.3). Basin's sides are 2000 m long, and depth is 30 m.

Experiments with only a few fluid elements are unrealistic because individual elements are discrete points and therefore cannot deform as an actual fluid does. Thus to be realistic, many fluid elements must be used. Our experiments document, however, that SEDCYC1 conserves energy and momentum, provided that many fluid elements are present in each cell.

ESTIMATING COEFFICIENTS OF FLOW

For realistic flow simulations, it is necessary to represent friction at the bottom and sides of channels and friction between adjacent portions of the fluid. Thus, if we deal with flow in open channels, we must obtain the coefficient of bottom friction c_1 and the coefficient of lateral friction c_2. There are suitable formulas for open-channel steady flow based on experimental data. The bottom-friction coefficient can be obtained when bottom friction is present and lateral friction is absent, as in the interior of a wide channel where the edges lie at appreciable distances. Under these circumstances, Equation 2.82 becomes

$$\frac{DQ}{Dt} = -g\nabla H - c_1\frac{Q|Q|}{H - Z} \tag{4.1}$$

Because the flow is steady and uniform, DQ/Dt is zero:

$$|Q| = \sqrt{\frac{g}{c_1}}\sqrt{Sh} \tag{4.2}$$

where

S = absolute value of slope = $|\nabla H|$
h = flow depth = $H - Z$

Equation 4.2 is similar to Chezy's equation:

$$Q_u = C\sqrt{R_h S} \tag{4.3}$$

where

Q_u = uniform flow velocity
C = constant
R_h = hydraulic radius (in a wide channel $R_h = h$)

If coefficient C in Chezy's formula can be determined, then Chezy's formula can be used to calibrate coefficient c_1 in Equation 4.2.

Coefficient of Bottom Friction

Manning (1890) proposed the following empirical relationship:

$$C = \frac{R_h^{1/6}}{n} \quad \text{(SI units)} \tag{4.4}$$

where n = Manning's roughness coefficient. Values for n are derived empirically and depend on the roughness of the material at the surface of the channel bed. Equating 4.2 with 4.3 and substituting 4.4, we obtain

$$c_1 = g\frac{n^2}{h^{1/3}} \quad \text{(SI units)} \tag{4.5}$$

which yields the bottom-friction coefficient.

Coefficient of Lateral Friction

The coefficient of lateral friction c_2 can be obtained with Manning's equation, which is derived from Equations 4.4 and 4.5:

$$Q_u = \frac{1}{n} S^{1/2} R_h^{2/3} \qquad \text{(SI units)} \tag{4.6}$$

Equation 4.6 is widely used for semiempirical representation of open-channel steady flow, and is useful for calibrating the coefficient of lateral friction c_2. If we assume a channel of rectangular cross section in which the flow is steady and "parallel" (i.e., there is no significant component of flow across the channel), the simplified two-dimensional momentum equation, Equation 2.81, is relatively simple to solve. Manning's equation also applies under these conditions, thereby allowing us to solve for c_2. However, if we solve for c_2, we find that c_2 takes slightly different values for channels of different geometric forms. Therefore it is appropriate to find a value of c_2 that produces the best overall agreement with Manning's equation for channels that vary over a range of slopes, widths, depths, and cross-sectional shapes.

Figure 4.6 shows calibration of c_2 in a channel of rectangular cross section that is 10 m deep and has a longitudinal slope of 0.01. The dashed curves in Figure 4.6 show average velocities as a function of channel width, and were calculated using Equation 2.81 and assuming steady "parallel" flow. Each curve was obtained with a different value of c_2. The bold line shows flow velocity as a function of width in accord with Manning's formula. For specific conditions of depth, slope, and roughness in Figure 4.6, a value of approximately 100 kg/(m s) for c_2 provides good agreement with Manning's formula over a wide range of channel widths. However, for optimum results, c_2 should be recalibrated for different conditions.

In deriving the value of the coefficient of lateral friction, c_2, we assume that there is no flow component across the channel. In SEDSIM, however, velocity components across channels are usually present, resulting in simulated flow velocities that differ slightly from those predicted under the assumption of steady parallel flow.

TESTING THE CALIBRATED FLOW MODEL

We tested SEDCYC1's performance under various flow conditions after it had been calibrated. The tests involved a trapezoidal channel 100 m wide at the base, with a slope of 1 in 100 (Fig. 4.7), identical to that used to calibrate c_2. The discharge rate was varied between 10 and 50,000 m^3/s. SEDCYC1's re-

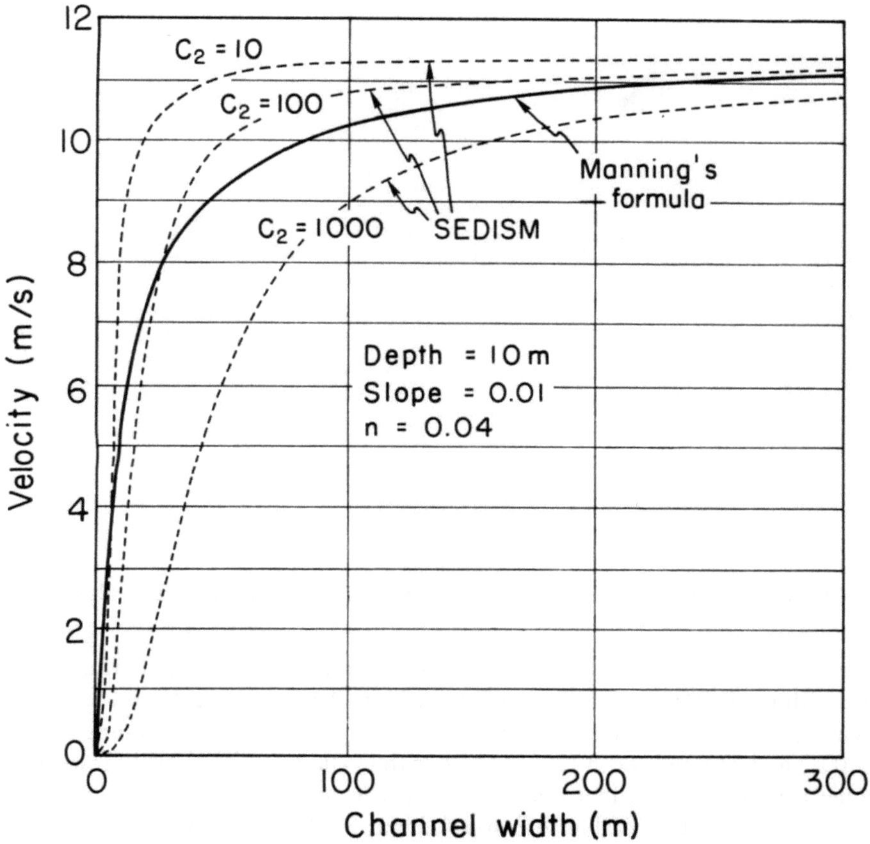

Figure 4.6 Functions representing average flow velocity versus channel width, for channels of rectangular cross section, that provide calibrations for lateral friction parameter c_2. Depth and slope are constant. Bold curve represents Manning's equation, whereas dashed lines are based on SEDSIM's equation for steady "parallel" flow for three values of c_2. Best match for Manning's curve ($c_2 = 100$) has been chosen for SEDSIM's three versions.

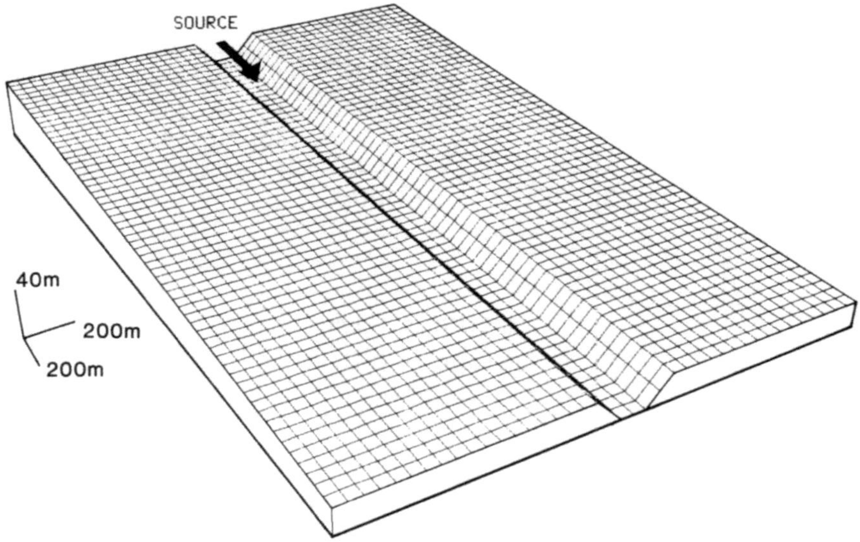

Figure 4.7 Channel used in experiment involving open-channel steady flow.

sponses were compared with predictions based on Manning's formula (Eq. 4.6) and with predictions stemming from application of Equations 2.25 and 2.81 for steady parallel flow. The experiments were repeated with a different slope (0.001) and channel width (200 m) to test the validity of SEDCYC1's response under conditions that differ from those under which the parameters were calibrated. The results in Tables 4.2 and 4.3 and Figure 4.8 show that velocities calculated by SEDCYC1 for given discharge rates are within 10% of those predicted by Manning's formula. This degree of accord may not be suitable for certain engineering applications, but when we consider that SEDSIM must be sufficiently versatile to simulate flow in irregular channels of various shapes, the accord is acceptable for our purposes.

CHANNEL-BEND EXPERIMENT

Having documented that SEDCYC1 is suitable for various straight channels, we then performed an experiment with a channel that bends. The initial channel is shown in Figure 4.9; its subsequent modification after a simulated day, in Figure 4.10. In the experiment, both ends of the channel were partly closed to keep the channel full of water, and the flow rate was adjusted to cause slight flooding of the channel's banks downstream. The vectors in

(Text continues on page 60.)

TABLE 4.2 Rates of Volumes of Water Discharged versus Fluid Velocities Obtained with Three Computationally Different Procedures (Data are plotted in Fig. 4.8.)

Discharge (m^3/s)	(*1*) *Velocity* (*Manning*) (*m/s*)	(*2*) *Velocity* (*steady parallel flow*) (*m/s*)	(*3*) *Velocity* (*turbulent flow*) (*m/s*)
10	0.689	0.678	
20	0.909	0.895	
50	1.309	1.28	
100	1.724	1.70	1.68
200	2.27	2.23	
500	3.26	3.21	
1,000	4.27	4.23	4.10
2,000	5.59	5.57	
5,000	7.90	7.99	
10,000	10.18	10.50	10.01
20,000	12.96	13.86	
50,000	17.35	19.82	

Note: Rectangular channel is 100 m wide, slope is 0.01, Manning's $n = 0.04$, and $c_2 = 100$ kg/(m s).

TABLE 4.3 Comparisons of Discharge Rates and Velocities Similar to Those of Table 4.2, Except That Channel is 200 m Wide and Slope is 0.001 (Data are Plotted in Fig. 4.8.)

Discharge (m^3/s)	(*1*) *Velocity* (*Manning*) (*m/s*)	(*2*) *Velocity* (*steady parallel flow*) (*m/s*)	(*3*) *Velocity* (*turbulent flow*) (*m/s*)
10	0.262	0.258	
20	0.346	0.340	
50	0.500	0.490	
100	0.656	0.645	0.611
200	0.864	0.849	
500	1.24	1.22	
1,000	1.63	1.61	1.53
2,000	2.14	2.11	
5,000	3.05	3.04	
10,000	3.96	3.99	3.45
20,000	5.10	5.24	
50,000	6.99	7.50	

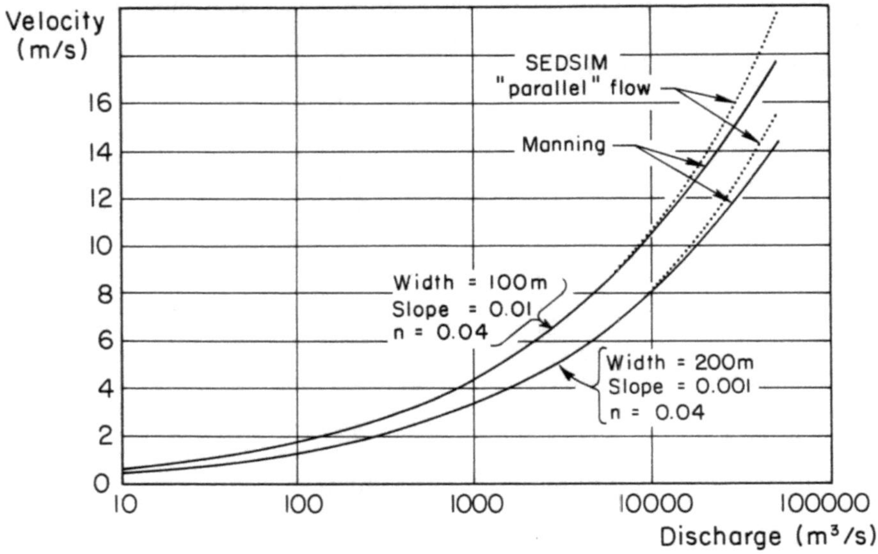

Figure 4.8 Plots of average flow velocity versus discharge rate in two open channels of differing slopes and widths. Data are listed in Tables 4.2 and 4.3. Curves (solid lines) based on Manning's equation are contrasted with results of SEDSIM's equation (dotted lines) for uniform parallel flow. Curves coincide at low values of discharge.

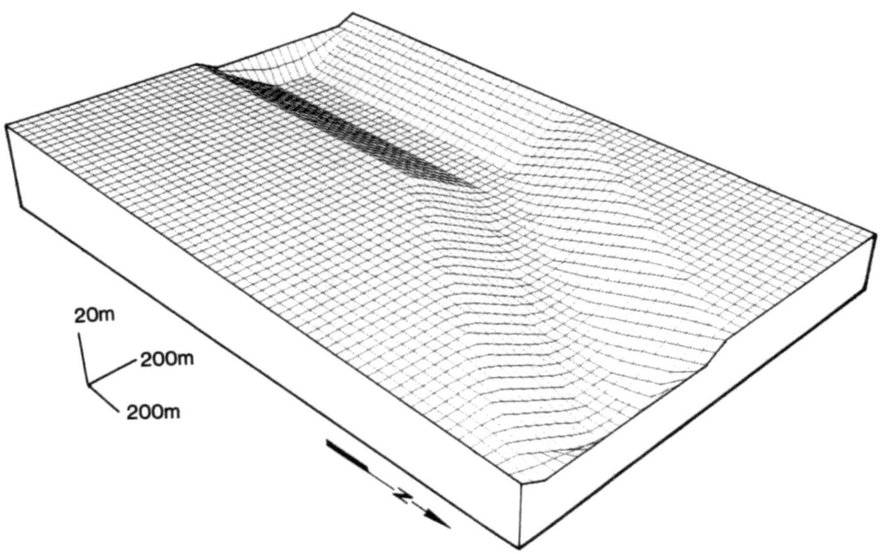

Figure 4.9 "Bent" channel that has trapezoidal cross section. Sills at ends of channel maintain substantial water depths so that flow velocities are reduced within channel.

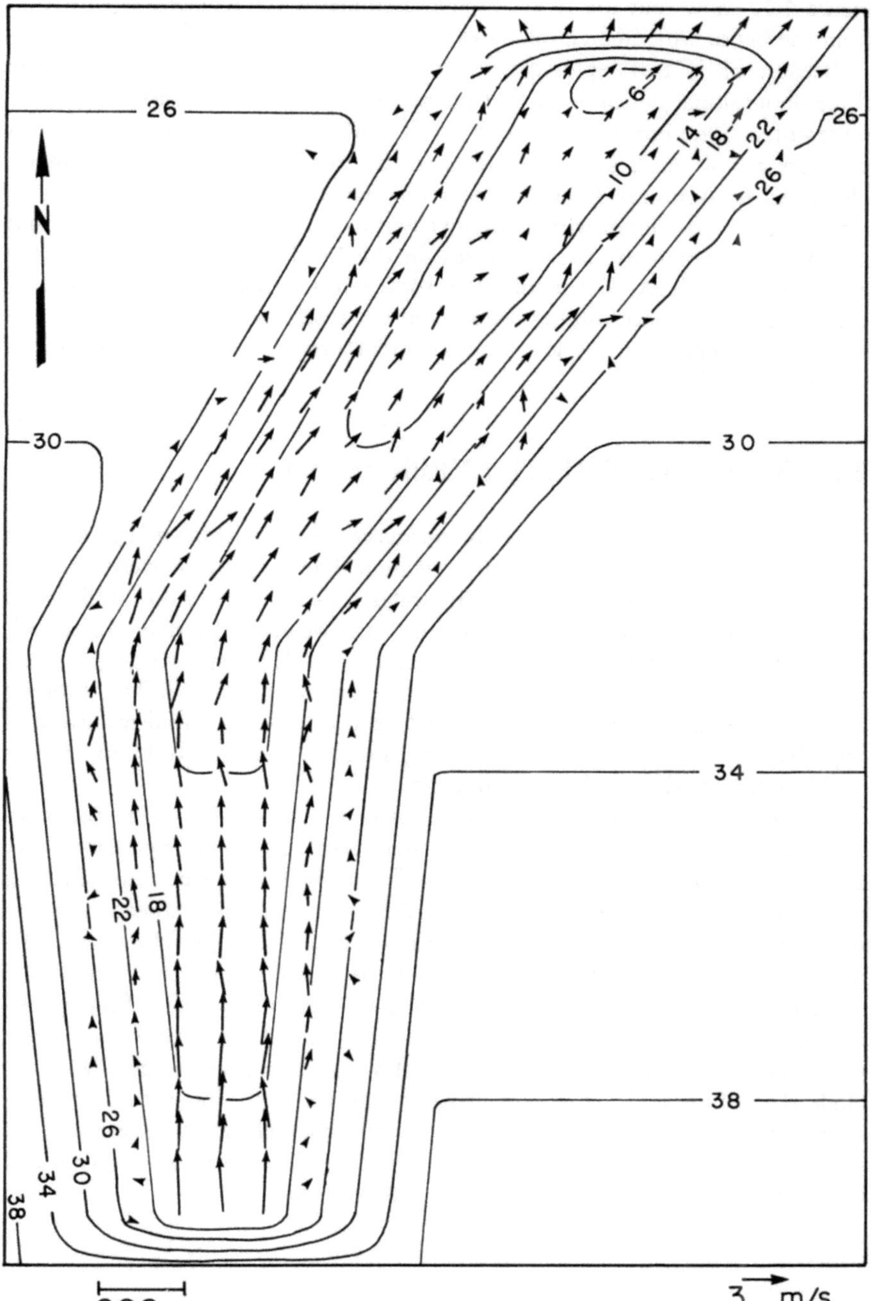

Figure 4.10 Map of bent channel shown in Figure 4.9, showing flow pattern after one simulated day. Arrows indicate flow velocities, and contours are in meters.

Figure 4.10 show that the flow becomes asymmetric immediately downstream from the bend, with higher velocities near the outside of the bend.

Because SEDCYC1 represents flow in only two horizontal dimensions, the three-dimensional helical flow that takes place as actual rivers meander is not represented. One consequence is that SEDSIM's representation of erosion on the outsides of bends and deposition on the insides is not wholly realistic. However, erosion and deposition at river bends are also affected by higher shear stresses associated with higher velocities on the outsides of bends and by lower velocities and shear stresses on the insides of the bends. Because SEDCYC2 and SEDCYC3 incorporate shear stresses, they respond by eroding the outside slopes and forming deposits on the inside slopes, although rates of erosion and deposition are less than in actual streams because helical flow is not represented.

TURBIDITY-CURRENT EXPERIMENT

A turbidity current is an unsteady subaqueous flow in which a large amount of relatively dense fluid is suddenly released. The turbidity current's greater density with respect to the surrounding water causes it to flow downhill. In an initial experiment involving a turbidity current, we created a steep surface adjoining a nearly horizontal lower surface (Fig. 4.11), representing a topographic situation similar in form to a continental slope and rise. The turbidity current was endowed with a density of 1.5, equivalent to a turbidity current of high density. Because SEDCYC1 can only simulate flow, the turbidity current could neither erode nor deposit sediment.

The results of the experiment are shown in a succession of vector maps (Figs. 4.12*a*–*f*) that represent the flow at intervals separated from each other by a minute. Early in the experiment the flow rushed downward until reaching the bottom of the steep slope (Figs. 4.12*b*), where it then spread over the nearly horizontal surface and slowed abruptly (Figs. 4.12*c*–*f*). The topography in the experiment is featureless except for the initial slopes, and it remained unchanged because erosion and deposition were prohibited. The flow moved in a relatively uniform path down the slopes, mimicking a turbidity current that debouches upon a gentle slope after flowing down a steep slope. We conclude that SEDCYC1's flow is consistent with the behavior of turbidity currents when conditions appropriate for turbidity flows are specified. In Chapter 6 we describe a similar turbidity flow that was permitted to erode and to deposit, which yielded very different flow patterns.

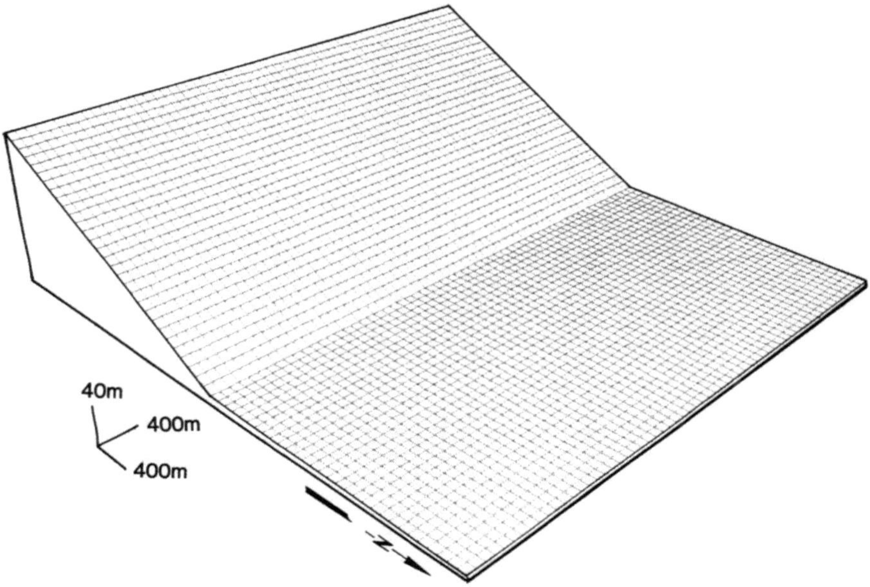

Figure 4.11 Submerged slope used in experiments involving high-density turbidity flows.

SUMMARY

The experiments described here document that SEDCYC1 can simulate flow over various topographic surfaces, with flows representing both subaerial streams and subaqueous turbidity currents. SEDCYC1 is limited, however, because it represents flow velocities in only two horizontal dimensions, although depth is represented so that the vertical dimension is partly represented, but helical flow is precluded. The other main limitation is that only flow profiles in which velocity is constant with depth are employed, because the computing effort would be too large otherwise. We conclude that although SEDCYC1 involves lengthy computations, it is flexible in simulating either steady or unsteady flow over irregular topographic surfaces, and it provides a foundation for flow models that transport clastic sediment, as described next.

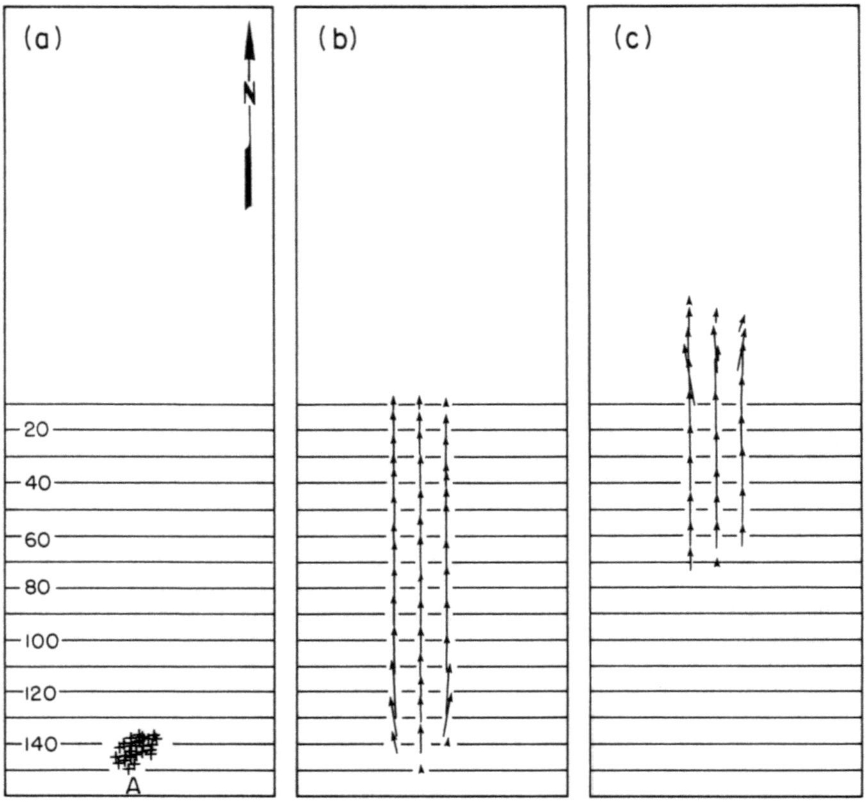

Figure 4.12 Contour maps of slope shown in Figure 4.11 in which flow patterns show path of high-density turbidity current. Arrows are proportional to velocities. Flow is released on steep slope at point A. Successive diagrams (b through f) show progress of flow at intervals of 1 minute simulated time. Contours represent elevation in meters above datum of zero at north edge of diagrams.

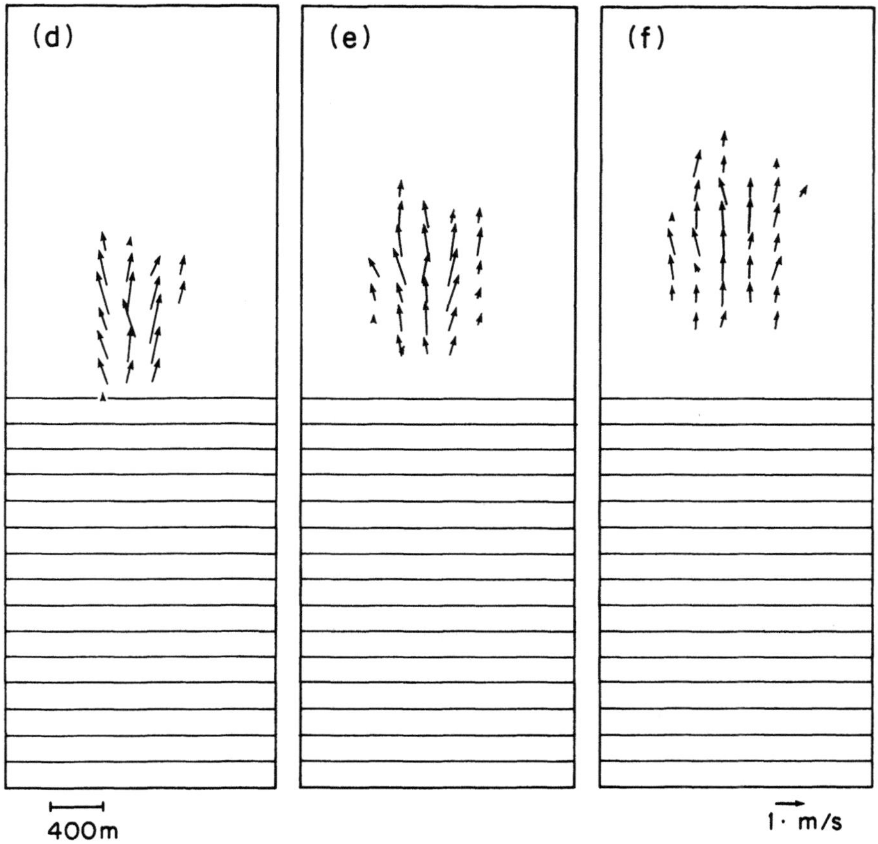

Figure 4.12 (continued)

CHAPTER 5

Background for SEDSIM2's Transport of a Single Type of Clastic Sediment

Chapters 2, 3, and 4 collectively dealt with the flow equations and their adaptation in SEDSIM1, which allows fluid elements to move in response to gravity, but does not allow them to erode, transport, or deposit sediment. If we extend the flow procedures so that they can transport sediment, the challenge is to determine how much sediment of specific grain size and density can be transported under prescribed flow conditions. Unfortunately, we lack comprehensive theory that relates erosion, transport, and deposition of particles to various conditions of flow, although we do have semiempirical formulas that are valid within specific ranges of particle sizes and flow conditions. Some of these formulas have been selected for incorporation in SEDSIM2 and SEDSIM3, yielding generalized and flexible procedures for eroding, transporting, and depositing clastic sediment.

THE SEDIMENT CONTINUITY EQUATION

When water flows over a topographic surface, the surface is lowered if erosion occurs and raised if deposition occurs. Thus, changes in topographic elevation due to erosion or deposition are necessarily linked with changes in the sediment content of the flow. During an increment of time, if erosion occurs and elevation is reduced, the amount of sediment added to the flow must equal the amount eroded. Similarly, when deposition occurs, the sediment carried by the flow must decrease and topographic elevation must increase. Thus, in simulation, accounts must be maintained to ensure that sediment is neither created nor destroyed, and that the "continuity" or conservation of material is scrupulously observed so that the models perform realistically. Furthermore, the accounting procedures must ensure that sediment be properly "credited" as it enters the system and "debited" as it exits.

We can represent the amount of sediment being transported by three alternative parameters, each of which can be converted to the others:

1. Sediment concentration expressed as mass or volume of sediment per unit volume of fluid
2. Sediment load expressed as concentration times flow depth
3. Sediment transport rate expressed as sediment load times sediment velocity

Continuity can be described by an equation that states that the amount of material eroded in an increment of time equals the sediment newly added to the flow, and that the amount of sediment deposited equals the amount subtracted from the flow:

$$(H - Z)\left(\frac{\partial l}{\partial t} + (\mathbf{Q} \cdot \nabla)l\right) = -\frac{\partial Z}{\partial t} \tag{5.1}$$

where

l = volumetric sediment concentration, which is defined as the volume of "settled" sediment per unit volume of fluid. A value of 0.01 for example, specifies that a unit volume of fluid (such as a cubic meter of water) contains an amount of sediment that, when completely settled, occupies a volume of 0.01 of the unit volume.
H = flow surface elevation with respect to sea level
Z = topographic elevation with respect to sea level
t = time
$\mathbf{Q}$ = horizontal flow velocity (defined by Equation 2.21)

Equation 5.1 can be written in Lagrangian form:

$$(H - Z)\frac{Dl}{Dt} = -\frac{\partial Z}{\partial t} \tag{5.2}$$

where D = derivative in the direction of flow.

Equations 5.1 and 5.2 imply that sediment moves at the same velocity as the flow. This is not strictly true in actual flows, because the total sediment load includes both *bed load* and *suspended load,* which move at different rates.

Particles in the bed load move slower because they are confined to a thin layer immediately above the stream bed, where they move by rolling, jumping (saltation), and dragging (traction). The suspended load, however, is maintained by turbulence and moves at approximately the same rate as the flow. In calculations, we can compensate for the slower movement of the bed load by reducing the proportion assigned to the bed load, a correction applied when flow models are calibrated.

TRANSPORTING SEDIMENT

The continuity equation ensures that sediment is conserved, but it does not prescribe how the flow conditions affect the amount of sediment transferred between the flow and the bed. The sediment load at a specific location and time depends on the sediment load previously supplied immediately upstream, plus momentary local changes in sediment load that depend upon flow depth and velocity, bed slope, and the slope of the flow's upper surface. Finally, the load also depends upon whether sediment or other material is available for erosion at the stream bed.

In SEDSIM2, variations in the load of sediment of a single grain size are expressed as

$$(H - Z)\frac{Dl}{Dt} = f(Q, \nabla H, \nabla Z, l, F) \tag{5.3}$$

where

f = load variation function
F = array of coefficients defining sediment properties $(f_1, f_2, \ldots, f_{N_f})$
N_f = number of coefficients

Equation 5.3 regulates the sediment load carried by the two-dimensional flow system of SEDSIM2. The function f in Equation 5.3 involves selection from commonly used formulas that pertain to sediment transport, as discussed later. The available formulas range widely in their applicability. Some relate only to threshold conditions for sediment movement, without specifying sediment load, whereas others apply only to bed load or to suspended load, and still others define sediment load under equilibrium conditions in which neither erosion nor deposition occur. SEDSIM2's procedures specify the total transport rate for sediment under equilibrium conditions, as well as rates of erosion or deposition when flow conditions change. SEDSIM2's procedures thus are generally compatible with widely used formulas for sediment transport and load.

FORMULAS FOR THRESHOLD MOVEMENT OF BED LOAD

Several criteria for sediment transport assume that particles of sediment begin to move when the flow's shear stress at the stream bed exceeds a value termed the critical shear stress. Shields' (1936) criterion assumes that the beginning of motion is a function of two dimensionless numbers: F_*, Shields' parameter, and R_*, the "particle Reynolds number," which are incorporated in Equations 5.4 and 5.5:

$$F_* = \frac{\tau_c}{(\gamma_s - \gamma)d} \tag{5.4}$$

$$R_* = \frac{v_* d}{\nu} \tag{5.5}$$

where

τ_c = critical boundary shear stress
γ_s = specific weight of sediment particles
γ = specific weight of water
d = sediment particle diameter
v_* = shear velocity = $\sqrt{\frac{\tau_0}{\rho}}$
ν = kinematic viscosity

Shields (1936) relates F_* and R_* as particles begin to move by expressing F_* as a function of R_*, as graphed in Figure 5.1.

The United States Bureau of Reclamation (Lane and Carlson, 1953) provides another method for calculating critical shear stresses needed to initiate movement of sediment particles. For particles with diameters larger than about 0.04 in or 1 mm, and that are not cohesive, the maximum shear stress in pounds per square foot is 0.4 times the particle diameter in inches. The shear stress is higher for finer particles. Figure 5.2 shows critical shear stress as a function of particle diameter, in accord with the Bureau of Reclamation's criteria. Figure 5.2 compares values for critical shear stress with values obtained with Shields' criterion, as well as with empirical values determined by Du Boys (1879). SEDSIM2 employs the Bureau of Reclamation's function graphed in Figure 5.2 for particles coarser than 0.1 mm, whereas for finer particles SEDSIM2 assumes that the maximum shear stress remains constant. This scheme is simple, yields suitably realistic results, and is easy to modify by adjusting a few parameters.

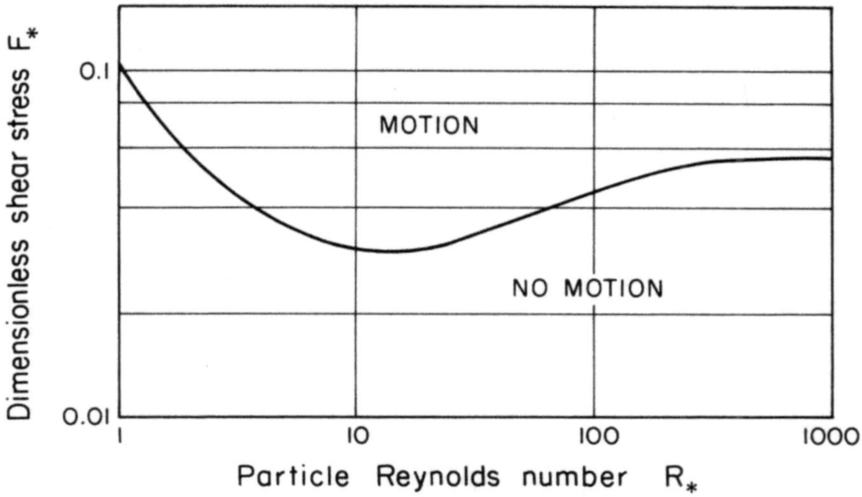

Figure 5.1 Shields's diagram showing dimensionless shear stress F_* as function of particle Reynolds number R_*. Values above curve represent conditions under which sediment is moved. After Shields (1936).

FORMULAS FOR TRANSPORTING PARTICLES AFTER THEY ARE SET IN MOTION

Because clastic sediment is transported both in suspension and in the bed load, formulas for both modes of transport are required. Du Boys (1879) provides a widely used formula for bed-load transport

$$Q_s = K\tau_0(\tau_0 - \tau_c) \tag{5.6}$$

where

Q_s = bed-load transport rate
K = transport coefficient
τ_0 = shear stress at the bed

Du Boys' formula and other bed-load formulas utilize the excess over the critical shear stress, $\tau_0 - \tau_c$, as the primary factor determining transport rate for bed load. The value of τ_c has been determined empirically by Du Boys (Fig. 5.2).

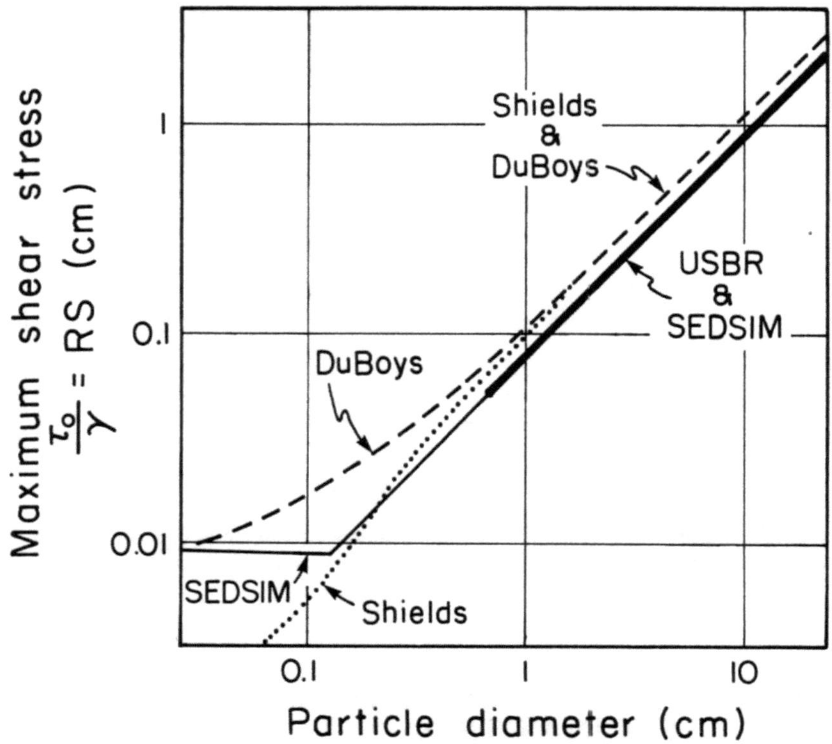

Figure 5.2 Maximum shear stress for cohesive and noncohesive materials as functions of grain size. Continuous line shows function employed by SEDSIM (bold portion of line coincides with function used by U.S. Bureau of Reclamation). After Lane and Carlson (U.S. Bureau of Reclamation, 1953), Du Boys (1879), and Shields (1936).

Meyer-Peter and Muller (1948) provide an alternative bed-load formula. They assume that part of the energy of the flow is dissipated in transporting sediment and that bed-load transport rate can be estimated from the loss of kinetic and potential energy of the flow. Their equation reduces to

$$Q_s = \frac{12.85}{\sqrt{\rho}\,\gamma_s}(\tau_o - \tau_c)^{3/2} \tag{5.7}$$

where

ρ = water density
γ_s = specific weight of sediment particles

SEDSIM2 employs a modified version of the Meyer-Peter and Muller formula (Eq. 5.7), because it can be adapted readily to calculate the total sediment load.

TRANSPORT FORMULAS EMPLOYED BY SEDSIM2

SEDSIM2 simulates the total sediment load consisting of bed load and suspended load. SEDSIM2's transport of sediment represents the balance between (1) the flow's transport capacity, Λ, which depends solely on flow conditions, and (2) the effective sediment concentration Λ_e. The effective sediment concentration is the volumetric sediment concentration l divided by factor f_1, called transportability, which reflects the ease with which the sediment is transported by the flow. If the effective sediment concentration is greater than the transport capacity, deposition occurs at a rate proportional to the excess effective concentration. If the sediment concentration is less than the transport capacity, erosion occurs if the critical shear stress is exceeded. Thus the transport equation may be written

$$(H - Z)\frac{Dl}{Dt} = \begin{cases} (\Lambda - \Lambda_e)f_2 & \text{if } \Lambda - \dfrac{l}{f_1} < 0 \text{ or } \tau_0 \geqslant f_3 \\ 0 & \text{if } \tau_0 < f_3 \text{ and } \Lambda - \dfrac{l}{f_1} \geqslant 0 \end{cases} \tag{5.8}$$

where

Λ = transport capacity (defined by Eq. 5.11)
$\Lambda_e = \dfrac{l}{f_1}$ = effective sediment concentration
f_1 = sediment transportability
f_2 = erosion-deposition coefficient
$f_3 = \tau_c$ = threshold shear stress for sediment movement

Equation 5.8 applies to subaerial streams and to turbidity currents. In hypopycnal flow, Equation 5.8 can only be used to represent deposition, because erosion does not occur. Shear stress at the bottom of the flow is given by

$$\tau_0 = c_1|\mathbf{Q}|^2\rho \tag{5.9}$$

where c_1 = bottom-friction coefficient utilized by Equation 2.80.

SEDSIM2's method for quantifying sediment transport resembles that of Meyer-Peter and Muller (1948). But SEDSIM2 assumes that power dissipated by bottom friction per unit volume of fluid, rather than energy loss per unit distance, controls transport capacity. This formulation agrees well with the concept of energy available for transport in sedimentary environments. Therefore

$$\Lambda = c_t \frac{dP}{dV} = c_t \tau_0 \frac{|Q|}{h} \tag{5.10}$$

where

P = power dissipated by bottom friction
V = volume
c_t = transport coefficient

Applying Equations 4.5 and 5.9 to 5.10 we obtain

$$\Lambda = c_t g \frac{n^2}{h^{4/3}} \rho |Q|^3 \tag{5.11}$$

where

g = gravity
n = Manning's roughness coefficient

Coefficients f_1, f_2, and f_3 in Equation 5.8 depend on sediment particle diameter, density, and shape. A procedure for calibrating them is described in Chapter 6. Transport capacity is described in the same units as volumetric sediment concentration l, which is dimensionless. Therefore the dimension of c_t is

$$[c_t] = \frac{\text{distance} \times \text{time}^3}{\text{mass}} \tag{5.12}$$

Because Equation 5.8 can be calibrated by adjusting coefficients f_1, f_2, and f_3, the value of c_t may be selected arbitrarily. For convenience, c_t is defined to be 1 ms^3/kg.

Equation 5.8 has proven satisfactory as a transport formula because it is sufficiently accurate for our purposes and can be readily represented by the particle-mesh method that we use. However, other formulas could be substituted in SEDSIM2 for Equation 5.8, if desired.

SEDSIM2'S TRANSPORT FORMULA COMPARED WITH OTHER FORMULAS

If Equation 5.8 is compared with other equations that represent sediment load and sediment transport, it is necessary to express these equations in the same general form as Equation 5.8. For example, Laursen (1956) modified several sediment transport equations to make transport rate depend on flow velocity, depth, and bed roughness. Table 5.1 shows Laursen's results, together with SEDSIM2's transport formula, reduced to the same form. Adaptation of SEDSIM2's formula in Table 5.1 involves the assumption that neither erosion nor deposition occurs, permitting the left side of Equation 5.8 to be set to zero, which, after applying Equation 5.12 and combining constants, yields

$$Q_s = B\frac{n^2}{h^{1/3}}|\mathbf{Q}|^4 \tag{5.13}$$

where B = constant.

Table 5.1 shows that SEDSIM2's formula resembles both Du Boys' formula and the formula of Meyer-Peter and Muller, in that transport rate is proportional to the fourth power of flow velocity. The main difference of SEDSIM2's formula's with respect to the other formulas in Table 5.1 is that SEDSIM2's formula assumes that the cube root of flow depth, rather than a higher power of flow depth, controls the transport rate. SEDSIM2's formula agrees better with experimental results for total load (Colby, 1964) than do the bed-load formulas presented in Table 5.1. Colby's results displayed in Figure 6.2 show that for high flow velocities, the total sediment transport rate remains constant or increases slightly with flow depth, rather than decreasing rapidly with depth as bed load does. Thus, according to Colby's data, SEDSIM2's formula, even though based on bed-load criteria, adequately represents total load.

Table 5.1 Comparison between Various Equations for Bed-Load Transport

Author	*Formula*
DuBoys (1879)	$Q_s = B_1 n^4 \frac{Q^4}{h^{2/3}}$
Meyer-Peter, Muller (1948)	$Q_s = B_2 n^3 \frac{Q^4}{h}$
Shields (1936)	$Q_s = B_3 \frac{n^4}{d} \frac{Q^5}{h^{2/3}}$
SEDSIM2 and SEDSIM3	$Q_s = B_4 \frac{n^2}{h^{1/3}} Q^4$

Q_s = sediment discharge rate
B_i = constant
n = Manning's roughness coefficient
d = sediment particle diameter
Q = water flow rate = $|\mathbf{Q}|$
h = flow depth

ESTABLISHING INITIAL CONDITIONS AND GEOGRAPHIC BOUNDARIES FOR OPERATING SEDSIM2

In addition to the initial conditions for flow (Eqs. 2.83 and 2.84), the initial sediment load must be prescribed when a simulation run is begun. The initial sediment load can be established by defining a function l_0 that assigns an initial sediment concentration at every point where flow occurs:

$$l(x, y, t_0) = l_0(x, y) \tag{5.14}$$

Topography, which was not allowed to change in SEDSIM1, must be prescribed as an initial condition in SEDSIM2. Boundary conditions that must be specified involve the flow boundary conditions provided by Equations 2.85 through 2.88, and also the sediment content at sources of flow and sediment content at the geographic boundaries. Sediment input at sources is represented by the sediment concentration of fluid elements when they enter the simulation area during each time increment. The sediment sources thus coincide with the flow sources. Sediment input can be specified throughout the simulation, as prescribed by the equation

$$l(x_{s_i}, y_{s_i}, t_0) = l_{s_i}(t) \tag{5.15}$$

where l_{s_i} = sediment concentration at source s_i.

Recall that the geographic boundary conditions specified for flow-only SEDSIM1 are defined so that the flow can exit (Eq. 2.87). SEDSIM2 (and SEDSIM3) use the same procedures, but constraints on erosion are needed. If flow occurs freely at boundaries and if erosion were unconstrained, all but the most cohesive materials would be rapidly eroded because a "waterfall effect" would ensue (Fig. 5.3*c*). Alternatively, if we closed the boundaries so that outflow is prohibited, water would be ponded in lakes and there would be excessive deposition near the boundaries where velocities are low (Fig. 5.3*a*). The compromise shown in Figure 5.3*b* is employed in SEDSIM2 and SEDSIM3. It permits flow to exit freely at boundaries but constrains erosion and deposition on the boundaries themselves while permitting erosion and deposition to occur elsewhere within the simulation area. A consequence is that deposits tend to wedge out within the zones adjacent to boundaries. The scheme is represented mathematically as

$$\frac{\partial Z(x_e, y_e)}{\partial t} = 0 \tag{5.16}$$

where (x_e, y_e) = points on the edge of the simulated area. Much as in a sand table, the effects imposed on sedimentation by the edges of the simulated area

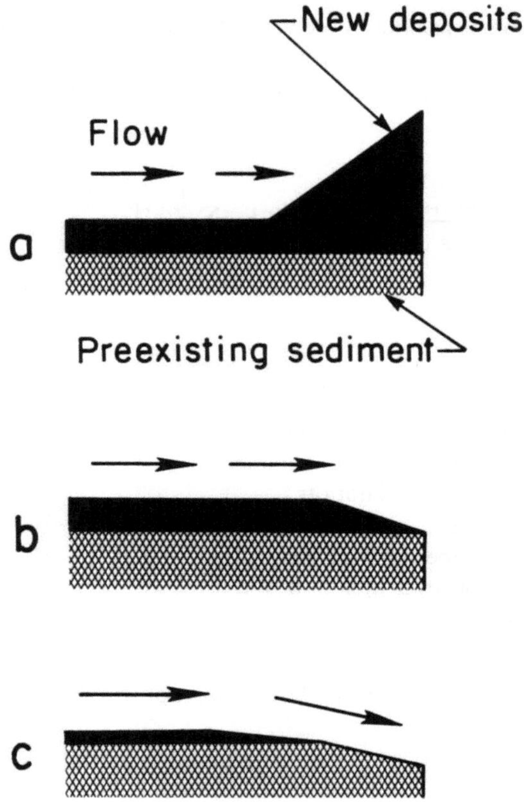

Figure 5.3 Three alternative procedures for treating boundaries in flow models. (a) Closed boundary prevents flow causing excessive deposition near boundary. (b) Open boundary permits flow of water and sediment across boundary, but erosion and deposition are prohibited in zone adjacent to boundary (procedure employed by SEDSIM). (c) Open boundary allows flow with no constraints on erosion or deposition, resulting in accelerated erosion.

cannot be wholly eliminated from a computer model, regardless of specific boundary schemes employed.

SEDSIM2'S PROCEDURE FOR APPROXIMATING THE SEDIMENT CONTINUITY AND SEDIMENT TRANSPORT EQUATIONS

The concentration of sediment in a moving volume of fluid at time t is equal to the initial concentration at time zero (Eq. 5.14) plus the net gain or loss of sediment between time zero and time t (integration of Eq. 5.8). Thus we can represent sediment concentration as a property of a fluid element that is updated at discrete time intervals by a numerical approximation of Equation 5.8.

Equation 5.2, the sediment continuity equation, is numerically approximated as follows:

$$\frac{Z_{i,j,t+1} - Z_{i,j,t}}{\Delta t} = -h_{i,j} \sum \frac{l_{k,t,+1} - l_{k,t}}{\Delta t} \tag{5.17}$$

where

Z = topographic elevation
i, j = row and column of node nearest to element k
t = index number of time increment
Δt = time increment duration
h = flow depth
k = index number of fluid element
Σ = summation over k, $1 \leqslant k \leqslant L$
L = number of elements near grid point (i, j)

Equation 5.17 can be used to update topographic elevation at every time increment:

$$Z_{i,j,t+1} = Z_{i,j,t} - Lh_e \sum \Delta l_k \tag{5.18}$$

where h_e = fluid element depth = V_e/A (Eq. 3.6).

Sediment transport is represented by Equation 5.8, with boundary conditions imposed by Equation 5.16, can be approximated as

$$h_e \frac{l_{k,t+1} - l_{k,t}}{\Delta t} = \begin{cases} \left(\Lambda_{i,j} - \dfrac{l_k}{f_1}\right) f_2 & \text{if } \Lambda_{i,j} - \dfrac{l_k}{f_1} < 0 \text{ or } \tau_{i,j} \geqslant f_3 \\ & \text{and } (i \neq 1, j \neq 1, i \neq N, j \neq M) \\ 0 & \text{otherwise} \end{cases} \tag{5.19}$$

where

N, M = number of rows and columns of grid nodes
$\tau_{i,j}$ and $\Lambda_{i,j}$ are defined as in Equations 5.9 and 5.11, but replacing Q by $Q_{i,j}$ and h by $h_{i,j}$.

At every time increment, the sediment concentration of each fluid element is updated as

$$l_{k,t+1} = l_{k,t} + C\frac{\Delta t}{h_e} \tag{5.20}$$

where C = right side of Equation 5.19. An intuitive explanation of the numerical method represented by Equations 5.18 and 5.20 is shown in Figure 5.4.

In summary, Equations 5.18 through 5.20 provide numerical approximations for transport of a single type of sediment, and, coupled with the flow model, they constitute the essence of SEDSIM2.

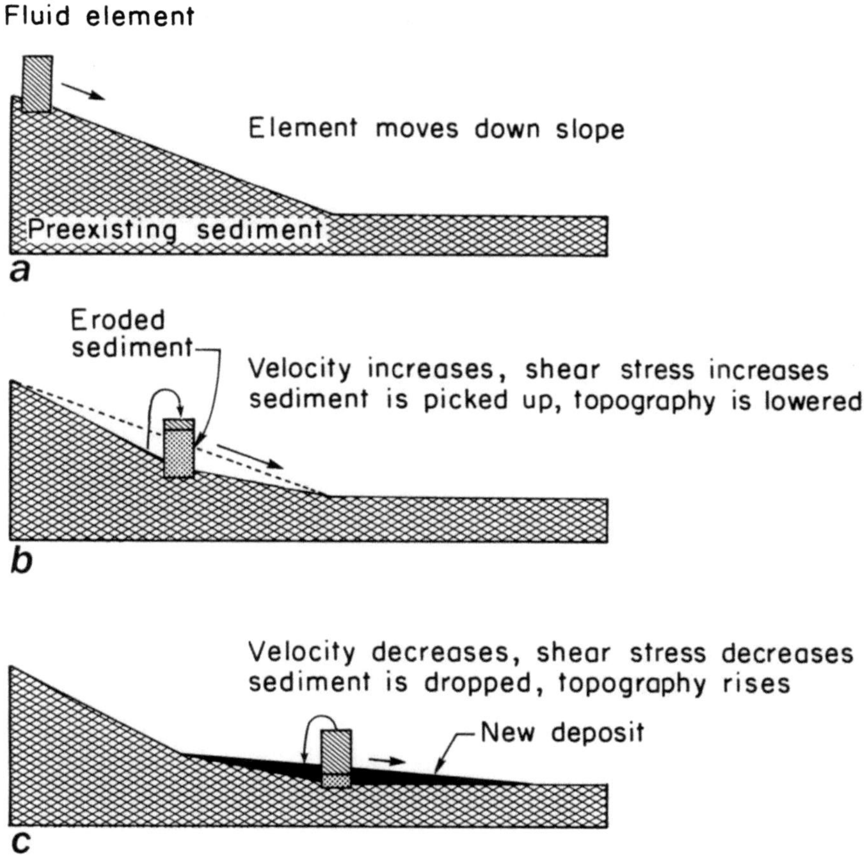

Figure 5.4 Mechanics of sediment transport employed by SEDCYC2 and SEDCYC3. (a) Fluid element moves down slope. (b) When velocity is sufficient, sediment is eroded and is transported in suspension, reducing topographic elevation. (c) When velocity decreases, reduced transport capacity causes sediment to be deposited, increasing elevation.

SUMMARY

We extended the flow model developed in the previous chapters by adding two equations that permit the model to simulate erosion, transport, and deposition of sediment of a single, uniform particle size. The two equations are (1) the sediment continuity equation (Eq. 5.2), which ensures that all sediment that is eroded, transported, or deposited is properly accounted for, and (2) the sediment transport equation (Eq. 5.8), which prescribes how much sediment of a given type can be transported by a flow of given hydraulic characteristics as well as the rate at which sediment is transferred between the flow and the underlying bed. The first equation is simple and should hold exactly under all flow conditions; the second equation is relatively complicated and is difficult to calibrate and test. Even though the sediment transport equation may only roughly approximate reality, our experiments show that it is sufficiently accurate for SEDSIM's purposes.

CHAPTER 6

Using SEDSIM2

SEDSIM2 builds upon SEDSIM1, but the tasks performed by SEDSIM2 are more extensive because they involve erosion, transport, and deposition of sediment as an aspect of flow. The flowchart for SEDSIM2 is identical to that of SEDSIM1 in Figure 4.1, but the tasks performed by each module are different. SEDCYC2 is the program that performs the simulation of flow and sedimentation. Input data files (Table 6.1) for SEDCYC2 are longer than for SEDCYC1 (Table 4.1) because information pertaining to sediment incorporated in fluid supplied at sources also must be provided.

SEDCYC2 accomplishes many additional tasks as compared with SEDCYC1, and these tasks are carried out by subroutine NEWELM2. NEWELM2 calculates the transport capacity of each fluid element during each computing cycle. If a fluid element's sediment concentration divided by its sediment transportability is greater than its transport capacity, deposition occurs. If it is less, erosion occurs, provided that the critical shear stress at the stream bed is equalled or exceeded (Eqs. 5.19 and 5.20). These calculations are made in program loops that update flow parameters during each computing cycle.

GRAPHIC DISPLAY

Graphic display of SEDCYC2's output involves program SEDSHO2, which is substantially longer than corresponding SEDSHO1. SEDSHO2 accomplishes more tasks. For example, SEDSHO2 updates the topographic contour map

TABLE 6.1 Example Input Data File for SEDCYC2

```
TITLE: SINGLE-SEDIMENT EROSION AND DEPOSITION ON INCLINE

RUN PARAMETERS:
START TIM=0.00E+00 Y
END TIM  =0.10E+00 Y
DISPL INT=0.10E+00 Y
TIM INCR1=0.50E+01 S
TIM FACT1=0.10E+04
TIM INCR2=0.10E+00 Y
TIM IDLE =0.00E+00 Y
TIM FACT2=0.10E+01

GENERAL PHYSICAL PARAMETERS:
FLOW DENS=    1000. KG/M3
SEA DENS =    1027. KG/M3
LAT FRICT=     100. NS/M2
ROUGHNESS=  0.0400
EL VOLUME=0.10E+05 M3
CURR(X,Y)=0.00E+000.00E+00M/S

SEDIMENT PARAMETERS:
                S1
DIAMETER = .50E-01 M
DENSITY  =   2700. KG/M3
BAS DECAY=   0.200
COHESION = .60E+00

SOURCES:
INTERVAL = .50E-04 Y
N SOURCES=         1
  X(M)     Y(M)    XV(M/S) YV(M/S)   S1(M3)
   1000.    2675.    -0.5     -1.5   .80E+04

TOPOGRAPHY:
GRID SIDE=   100.0 M
NROWS    =         5
NCOLS    =         5

GRID NODES ELEVATION (SURFACE) (M)
   50.0   50.0   50.0   50.0   50.0
   45.0   45.0   45.0   45.0   45.0
   40.0   40.0   40.0   40.0   40.0
   35.0   35.0   35.0   35.0   35.0
   30.0   30.0   30.0   30.0   30.0

ELEMENT POSITIONS AND VELOCITIES:
N OF ELEM=         0
  X(M)     Y(M)    XV(M/S) YV(M/S)   S1(M3)
```

each time a new display is desired because the topography changes during each program cycle. Thus, at intervals specified by the user, subroutine WRITGU2 writes new topographic values to the graphics file.

SEDSHO2 also draws vertical sections that show the ages of deposits and their thicknesses. The actual drawing is handled by subroutine CRSSEC2. To construct a section, CRSSEC2 asks the user for a geographic point and an azimuth direction to define the trace of the section. During each display cycle, CRSSEC2 draws the topographic surface intersected by the section. Where the topographic surface is higher than the surface in the previous display, new deposits have formed. The sediment deposited during each time increment is then labeled with a color that specifies its age (the user specifies the number of years in each increment). If erosion reduces the elevation, the eroded deposits are erased from the section.

ECONOMIZING ON ARITHMETIC

Simulating processes through geologic time provides a major computing challenge because time increments as short as a few seconds are necessary for realistic flow simulation. We must reconcile the need to represent geologic time over tens of thousands of years, with the need to represent changes in flow in intervals measured in seconds. The challenge is to economize on arithmetic and yet satisfy these demands.

We might attempt to speed up flow so that erosion, transportation, and deposition operate faster, thus in effect speeding up geologic time. For example, we could speed sedimentary processes simply by increasing transport coefficient c_t. Unfortunately, excessively high values of c_t lead to unrealistic results, so that simple "speeding up" is not satisfactory.

Compute-and-Drift Schemes

An alternative is to operate on a "compute-and-drift" basis. In many natural environments, the processes of erosion, transportation, and deposition occur slowly and steadily. This suggests that we might simulate sedimentary processes for only a short period (say for a few days), and then simply "drift" by extrapolating the simulation for a longer period, such as a month or a year. Such a compute-and-drift cycle could be repeated indefinitely. SEDCYC2 (and SEDCYC3) can "remember" the immediately previous topographic grid, and can then operate for a few days of simulated time, producing a new topographic grid that can be compared with the previous one. The differences between the two grids then can be multiplied by a factor that represents an extrapolation over a much longer interval. For example, if during a simulated day, 1 mm of sediment is eroded at a given location, a year's equivalent would be 365 mm. Such a scheme is computationally simple, but it can be carried only so far before flow and topography are seriously out of equilibrium. Thus, after a certain amount of erosion and deposition have occurred, it is necessary to recalculate the flow because it will no longer be in equilibrium with the topography.

It is essential that flows be in close accord with topography because of their close interdependence. Extrapolations that are too long may result in unrealistic holes and mounds where the flow is no longer appropriate, particularly near sources where fluid and sediment are supplied, and erosion and deposition tend to be rapid. However, if compute intervals and drift intervals are carefully selected, compute-and-drift schemes work well and greatly reduce the overall computational effort.

Compute-and-Stop Schemes

Compute-and-drift schemes are not appropriate for highly unsteady flows, such as turbidity currents or floods, which vary continually and never stabilize. Fortunately, highly unsteady flows are usually of short duration and are separated by long periods of relative inactivity. This permits us to confine calculations to brief intervals when the flow is unsteady, and then simply stop between calculations. For example, a turbidity current may flow down a submarine canyon for minutes or hours, and months or years may elapse before the next turbidity flow occurs. Or a river's annual flood may last only a few days, accomplishing virtually all of the erosional, transportational, and depositional activity for the entire year. Thus, we can simulate an unsteady flow and then proceed directly to the next unsteady flow, ignoring the time between, thus in effect computing and stopping. Such "compute-and-stop" schemes permit us to span intervals of geologic time by entirely confining the computing effort to the infrequent unsteady flows.

VERIFYING SEDSIM2

SEDSIM2 can be tested with experiments much as an experimental sand table might be used. The tests involve simple topographic forms created by "molding" channels or slopes from sediment whose properties are specified. Then water and sediment are supplied at specific rates, and the results are observed.

In our experiments we first calibrated coefficients that control sedimentary processes within SEDSIM2, such as sediment transportability. The coefficient of sediment transportability, f_1, in Equation 5.8, governs the amount of sediment that can be carried in the flow under equilibrium conditions, when there is neither erosion nor deposition. Fortunately, experimental data are available for calibrating f_1, because we need different values for f_1 depending on the size, shape, and density of particles of sediment. These characteristics can be represented by their aggregate effect upon a particle's velocity as it falls through water. As an approximation, we can assume that f_1 is inversely proportional to fall velocity:

$$f_1 = \frac{c_f}{W} \tag{6.1}$$

where

c_f = constant of proportionality

W = particle fall velocity

Under the condition specified by Equation 6.1, it is necessary to find only a single value of c_f to calibrate f_1 for any sediment whose fall velocity is known.

Most data relate sediment transport to hydraulic parameters of the flow and to grain size rather than to grain fall velocity, which requires that we convert grain sizes to fall velocities. By equating weight and drag, we compute fall velocity by

$$W = \sqrt{\frac{\gamma_s - \gamma}{\gamma} \frac{4}{3} \frac{d}{C_d}} \tag{6.2}$$

where C_d = drag coefficient. Empirical values of C_d as a function of Reynolds number for spheres are graphed in Figure 6.1. SEDSIM2 estimates fall velocity by using drag-coefficient values, indicated by the bold line in Figure 6.1. Within the range of Reynolds numbers where drag coefficient remains constant at 0.55, Equation 6.2 yields fall velocity. Where the drag coefficient is represented by an inclined straight line in Figure 6.1, the following formula provides fall velocity:

$$W = \left(\frac{\gamma_s - \gamma}{\gamma} g \frac{4}{90}\right)^{4/5} \nu^{3/5} d^{7/5} \qquad \text{(SI units)} \tag{6.3}$$

where ν = kinematic viscosity of water. Equations 6.2 and 6.3 can be combined for estimating particle fall velocities in water at 15°C:

$$W = \min(1895 r_\gamma^{4/5} d^{7/5}, 4.88 r_\gamma^{1/2} d^{1/2}) \qquad \text{(SI units)} \tag{6.4}$$

where $r_\gamma = (\gamma_s - \gamma)/\gamma$.

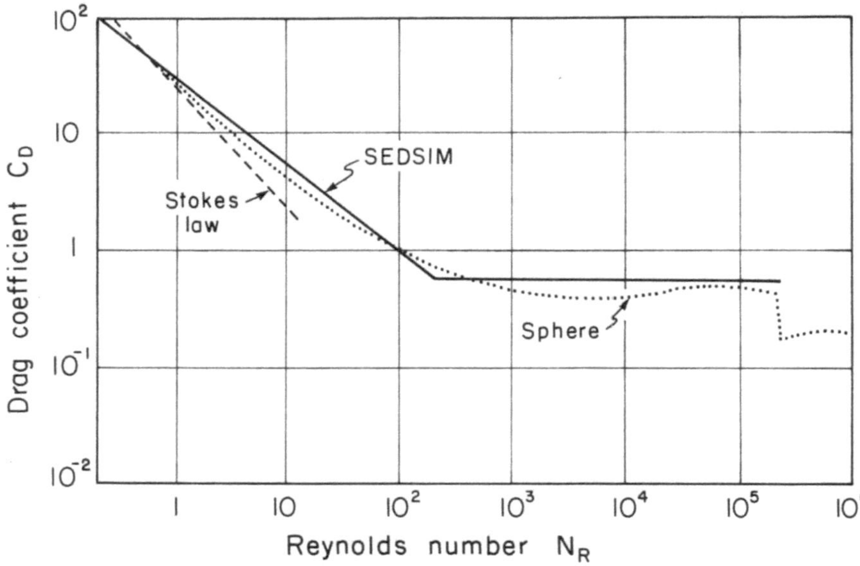

Figure 6.1 Drag coefficients as functions of Reynolds number for spheres (after Prandtl, 1930), and for Stokes law. Bold line indicates drag-coefficient function employed by SEDSIM for calculation of fall velocities of sediment particles.

Given Equations 6.1 and 6.4, it is possible to find a value for c_f that agrees best with empirical data for various types of sediment that relate sediment load or transport rate to conditions of flow under equilibrium conditions. Data for transport rates for sands (Colby, 1964) are illustrated in Figure 6.2. A value for c_f of 1.18×10^{-6} s/m best fits the data for sands.

When erosion or deposition occur, the rate at which sediment is transferred between the flow and deposits at the base of the flow is governed by coefficient f_2. If f_2 is large, the transfer rate is high, and if f_2 is infinite, the transfer is instantaneous. It is difficult to calibrate f_2 because data pertaining to transfer rates are sparse, but different values of f_2 have little effect on SEDCYC's performance. Most formulas that deal with sediment load or transport rate either assume that the sediment load adjusts instantaneously to changing hydraulic conditions, or else they ignore the rate of adjustment.

We use the same value of f_2 for both erosion and deposition. During deposition we assume that f_2 depends on the time it takes the sediment to settle out of the flow, the time being closely related to fall velocity. During erosion, we assume that fine sediment with a lower fall velocity takes longer to achieve equilibrium than does coarse sediment. For lack of adequate observational data, and because f_2 must be expressed in the same units as velocity, we assume that f_2 is equal to fall velocity W. Thus, if a flow that is in equilibrium with the sediment it carries suddenly comes to rest, SEDSIM2 causes the flow's sedi-

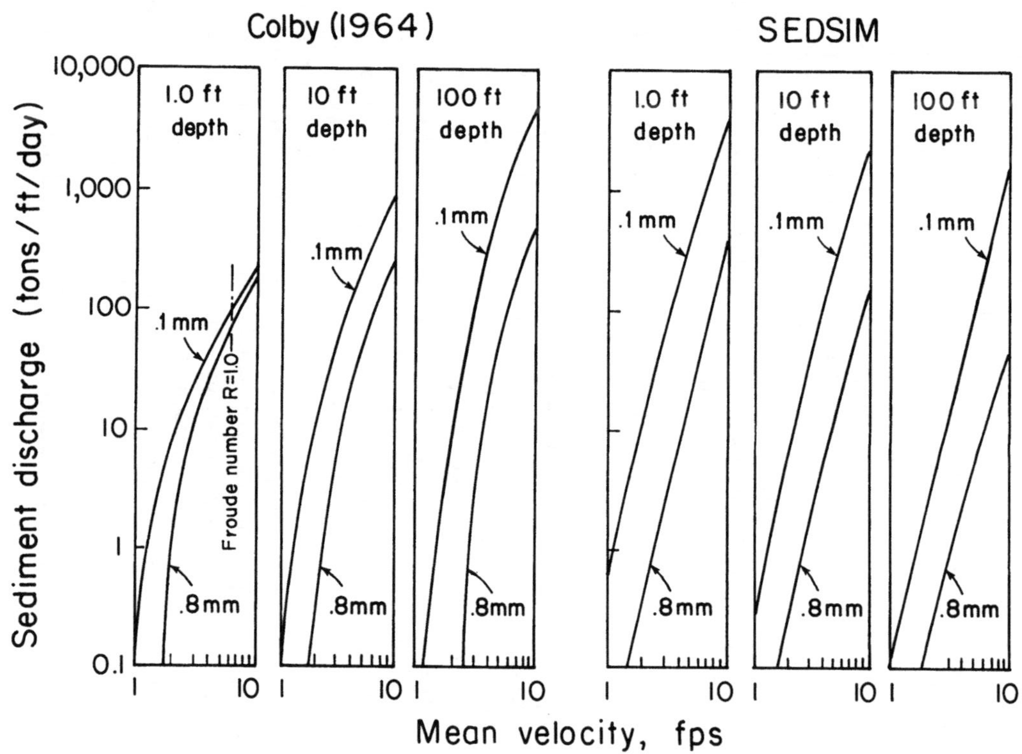

Figure 6.2 Relationship between discharge of sands to mean velocities for two sizes of bed sands and three depths of flow, based on observational data provided by Colby (1964) and as calculated by SEDSIM.

ment concentration to diminish initially at the same rate that actual sediment would diminish if it were uniformly distributed throughout the flow.

Coefficient f_3 is the critical shear stress, which SEDSIM2 calculates using a modified version of the U.S. Bureau of Reclamation's method (Lane and Carlson, 1953) described previously.

INITIAL EXPERIMENTS WITH SEDSIM2

After SEDSIM2's coefficients were calibrated, we conducted experiments to observe SEDSIM2's performance in different sedimentary environments. The first set of experiments involves comparison with the Colorado River and the Niobrara River, employing data provided by Vanoni and Nomicos (1960) and shown in Figure 6.3, which relates flow rate and sediment discharge for the two rivers and compares actual measurements with values predicted with various sediment-discharge formulas.

We conducted three experiments with SEDSIM2 for comparison with data from the two rivers. We assumed an initial topographic configuration similar to Figure 4.7. We assumed various flow rates at fluid sources placed near the upper end of the channel, and then observed the sediment discharge rates at the lower end of the channel. SEDSIM2's values are represented by crosses in Figure 6.3, and are in general agreement with actual values, although the actual values are more scattered, probably in response to variations in processes that SEDSIM2 does not represent adequately in its relative simplicity.

CHANNEL-BEND EXPERIMENT

We tested SEDSIM2 with the channel-bend experiment used initially with flow-only program SEDSIM1 and described in Chapter 4. The initial topography was the same, involving a channel containing a bend (Fig. 4.9), but now both erosion and deposition are represented. We assumed that the area contained loose homogeneous fine sand and that the flow rate was sufficient to cause the channel to overflow slightly. The experiment involved a simulated time of one year, and initial results are shown in Figures 6.4 and 6.5. After 0.2 years of simulated time, there has been some erosion in the channel, causing underwater dunes to form (Fig. 6.4). The average "wavelength" or distance between dunes is a small multiple of the cell size used in the grid, and we are not sure whether the average distance between dunes is in accord with flow conditions, or whether it has been influenced by cell size. Nevertheless, the flow conditions in the experiment cause underwater dunes to be created. For example, erosion occurs on the upstream faces of dunes in response to de-

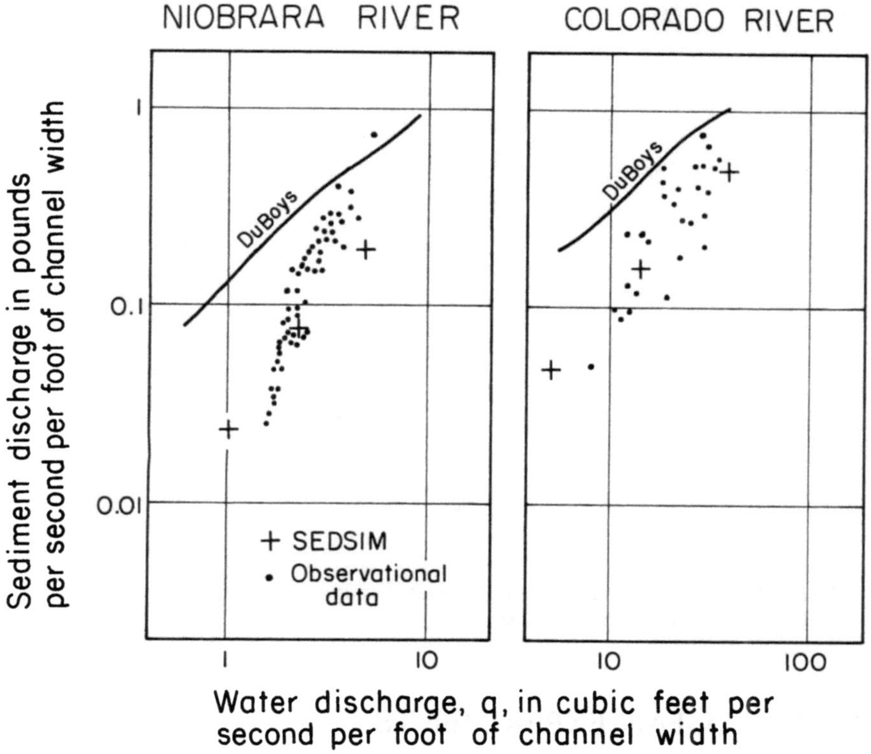

Figure 6.3 Sediment discharge as function of water discharge. Dots indicate values observed in natural streams (adapted from Vanoni and Nomicos, 1960). Crosses indicate values predicted by SEDSIM using straight channel shown in Figure 4.7. Solid lines indicate relationships calculated with criteria provided by DuBoys (1890).

creases in depth and increases in flow velocity, whereas flow velocity decreases on the downstream faces, causing deposition to occur. Experiments with a finer grid should reveal whether the rates of erosion and deposition on the dune faces are realistic.

After a simulated year (Figs. 6.6, 6.7, and 6.8 on pages 90, 91, and 92), substantial erosion has occurred near the fluid sources and the channel bottom has become increasingly irregular. The channel's cross section has also changed. Levees have formed where overbank flooding occurred, and enlargements of section B-B′ (Fig. 6.8) reveal the details of the levees and the alluvial plain beyond. SEDSIM2 appears to be effective in representing flow in channels, but SEDSIM2 could be extended to perform more realistically.

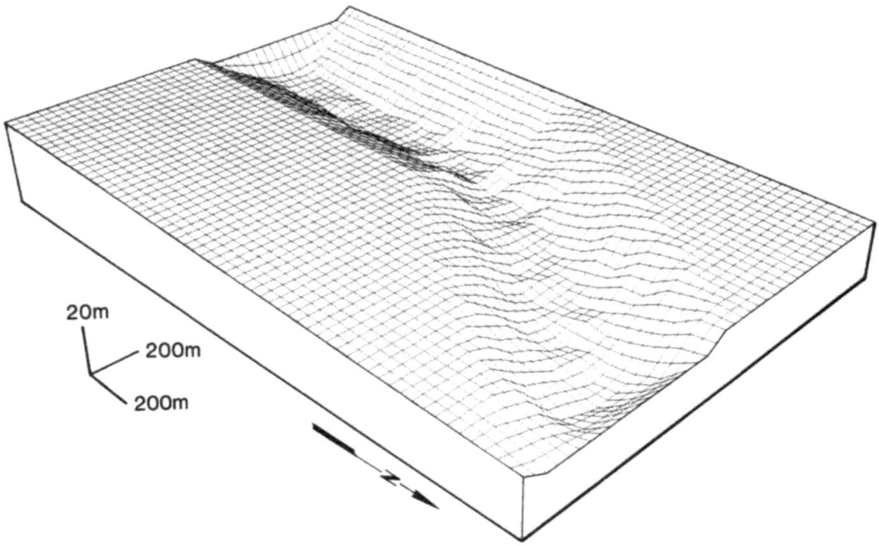

Figure 6.4 Bent channel after one simulated year. Medium sand forms channel and is available for transport by stream.

ALLUVIAL-FAN EXPERIMENTS

SEDSIM2 effectively simulates deposition of alluvial fans. When a fan is formed, the proportion of fluid with respect to sediment is relatively small, which is a computational advantage because the number of fluid elements can be reduced, decreasing computing time. We conducted two experiments involving alluvial fans, both with the same initial topography in a rectangular area 2 × 3 km and containing a subaerial surface that slopes gently except where broken by a steep scarp (Fig. 6.9 on page 93). The sediment at the surface is poorly consolidated sand of uniform grain size.

The flow comes from two separate sources. The resulting "snapshots" (Figs. 6.10 and 6.11 on pages 94–97) show that two main valleys developed on the steep slope, with corresponding alluvial fans formed where the slope decreases. There are also small valleys near the main valleys. One of the small valleys (marked X in Figs. 6.11*b* and *c*) was traversed by a stream for about the first 40 years of the experiment, but was abandoned later (Fig. 6.11*c*) when the stream was "captured" upstream.

Fishnet displays of fans at 20 and 100 years simulated time are shown in Figures 6.10*a* and *b*. The geologic section (Fig. 6.12 on page 98) shows sediment present when the experiment began with a crosshatched pattern. Newly formed deposits are shown bounded by time lines drawn every 20 years, in effect labeling deposits according to age.

(Text continues on page 93.)

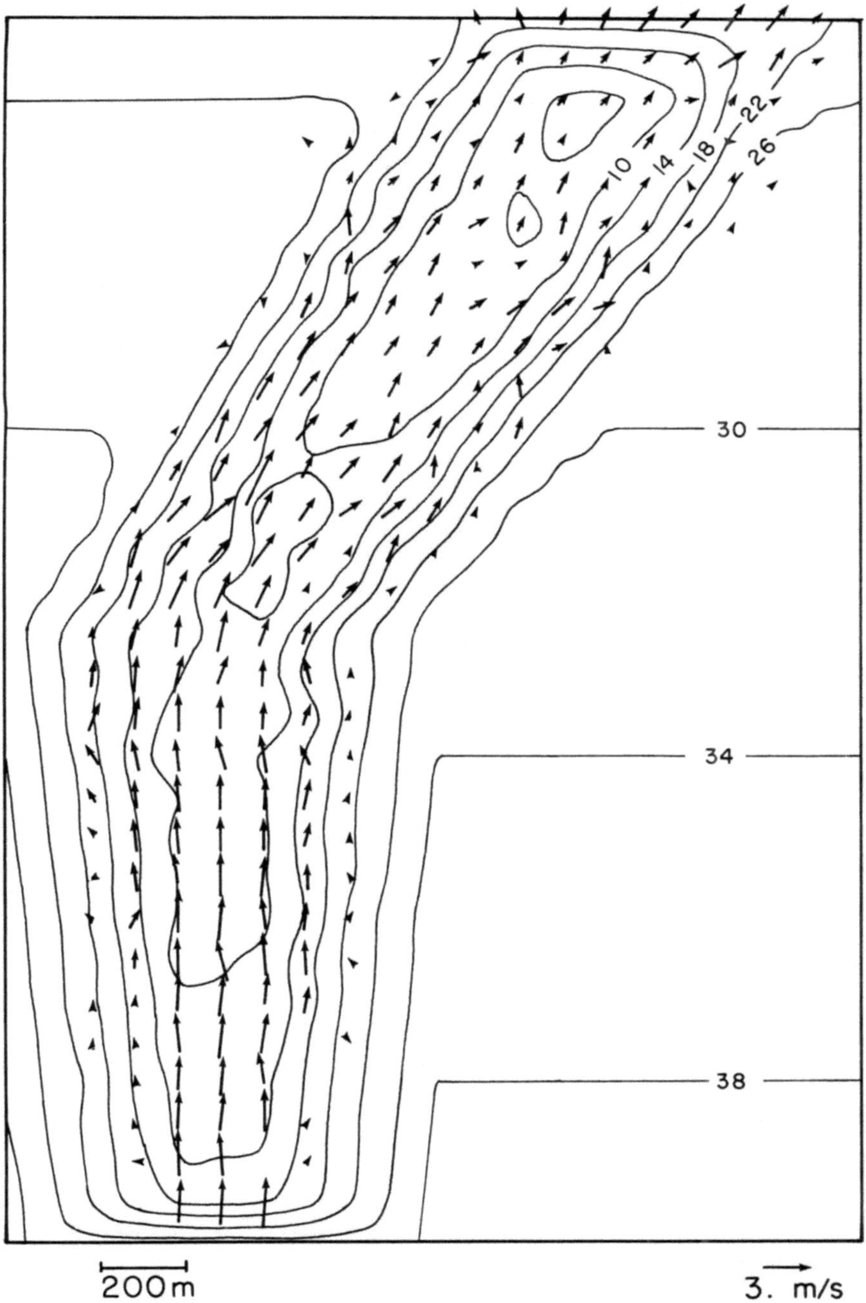

Figure 6.5 Map of bent channel showing contours and flow velocities after one simulated year.

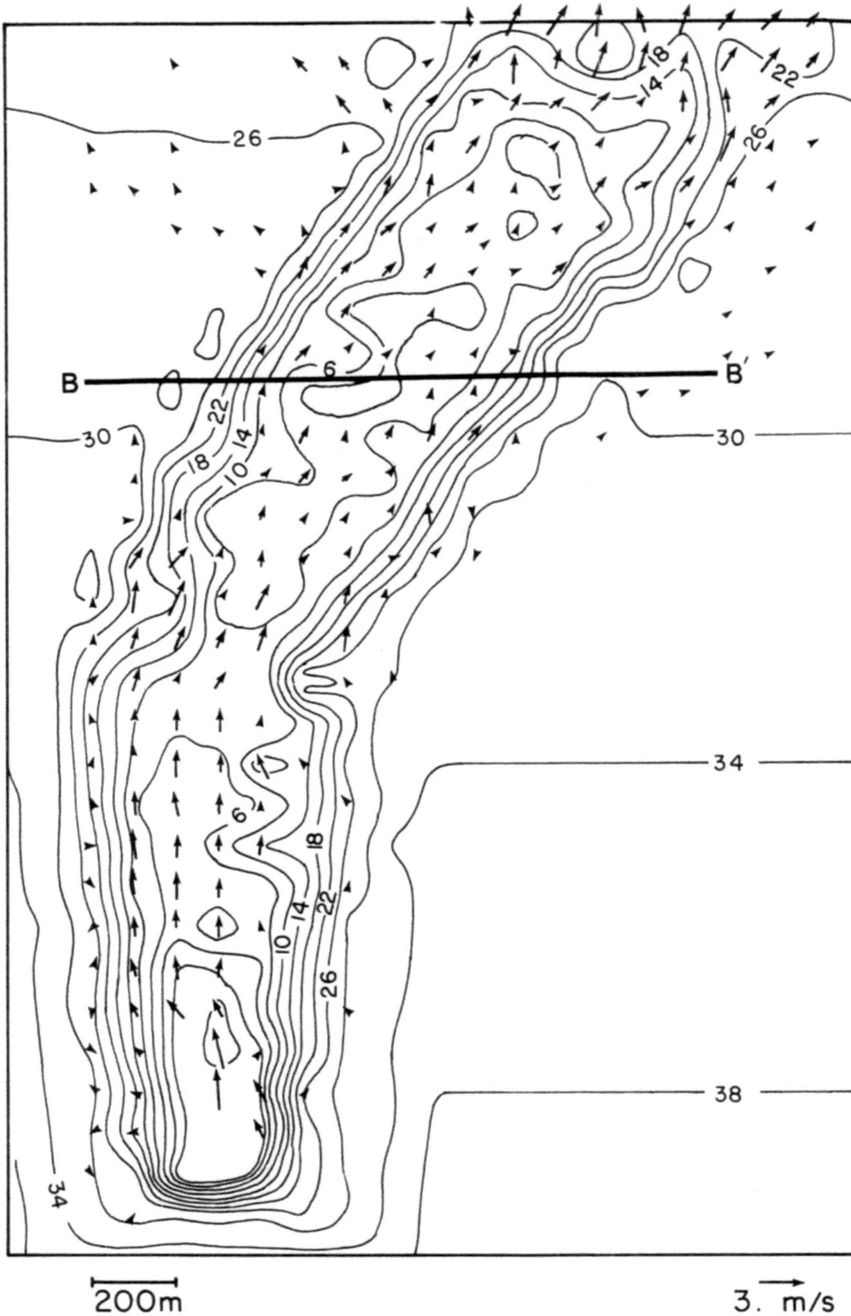

Figure 6.6 Contour map of bent channel after five simulated years. Section *BB'* is shown in Figure 6.7.

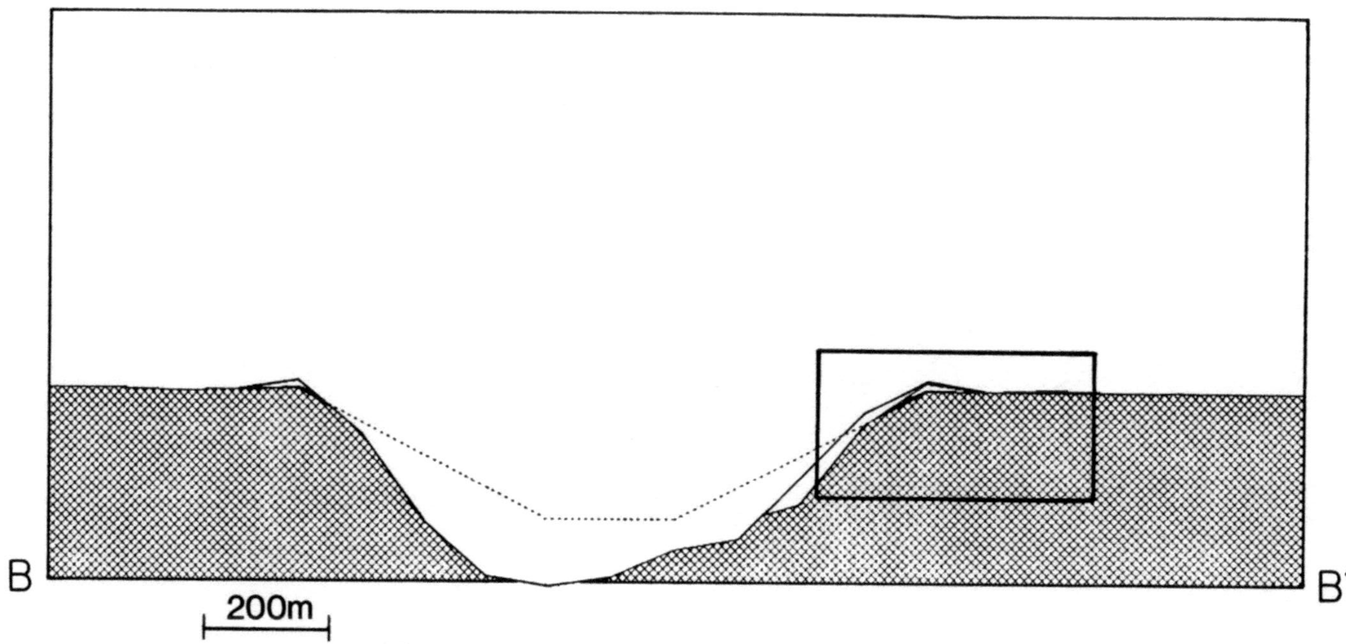

Figure 6.7 Section B-B′ (Figure 6.6) after five simulated years. Dotted line shows original channel. Levees and alluvial plain deposits have formed on both banks. Box outlines enlarged portion in Figure 6.8.

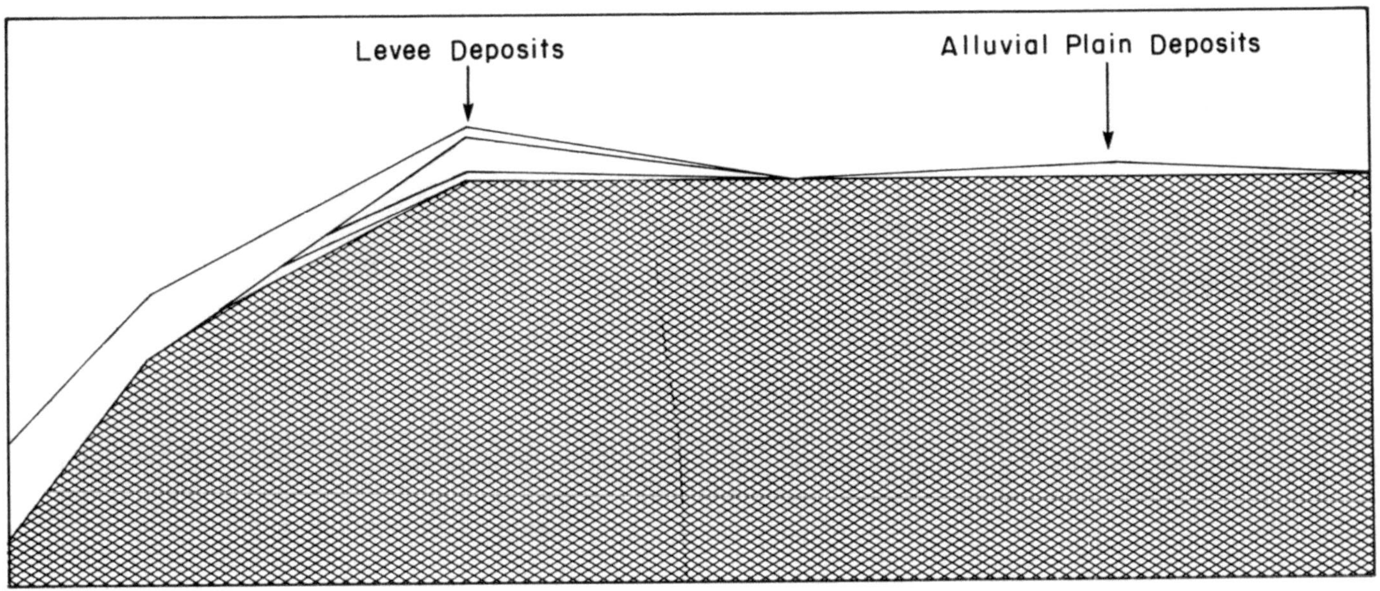

Figure 6.8 Enlargement of part of section B-B′ in Figure 6.7, showing details of levee and alluvial plain deposits.

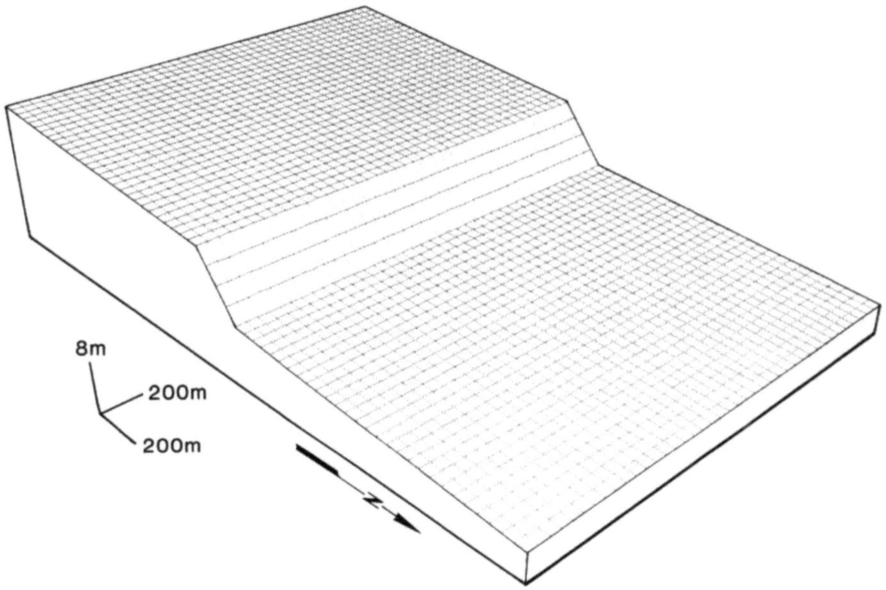

Figure 6.9 Initial topography in alluvial-fan experiment. Vertical exaggeration is 25.

The longitudinal section in Figures 6.12*a* and *b* shows that the long profile of the valley and fan approximate an exponential curve, in general accord with an equilibrium stream profile. Figure 6.12*b* enlarges part of the longitudinal section to reveal details of the new deposits. Transverse cross sections (not shown) reveal that areas of maximum deposition have shifted laterally.

Although the initial topography is starkly simple in these experiments, the flow patterns and the subsequent topography are irregular. The irregularities arise even though SEDSIM2 contains no random components. Such behavior is realistic because natural flows tend to behave randomly or chaotically, as discussed in Chapter 1.

TURBIDITY-CURRENT EXPERIMENT

We used SEDSIM2 to simulate turbidity currents, much as described in Chapter 4, except that erosion, transport, and deposition of sediment are represented by SEDSIM2. Each turbidity current was initiated by releasing 2000 m^3 of water containing 1000 kg of fine sand per cubic meter of water at the location shown in Figure 6.13*a* (page 100). After release, the relatively dense mass of turbid fluid glided down the slope, its progress being shown in Figures 6.13*b* through *f* (pages 100–101).

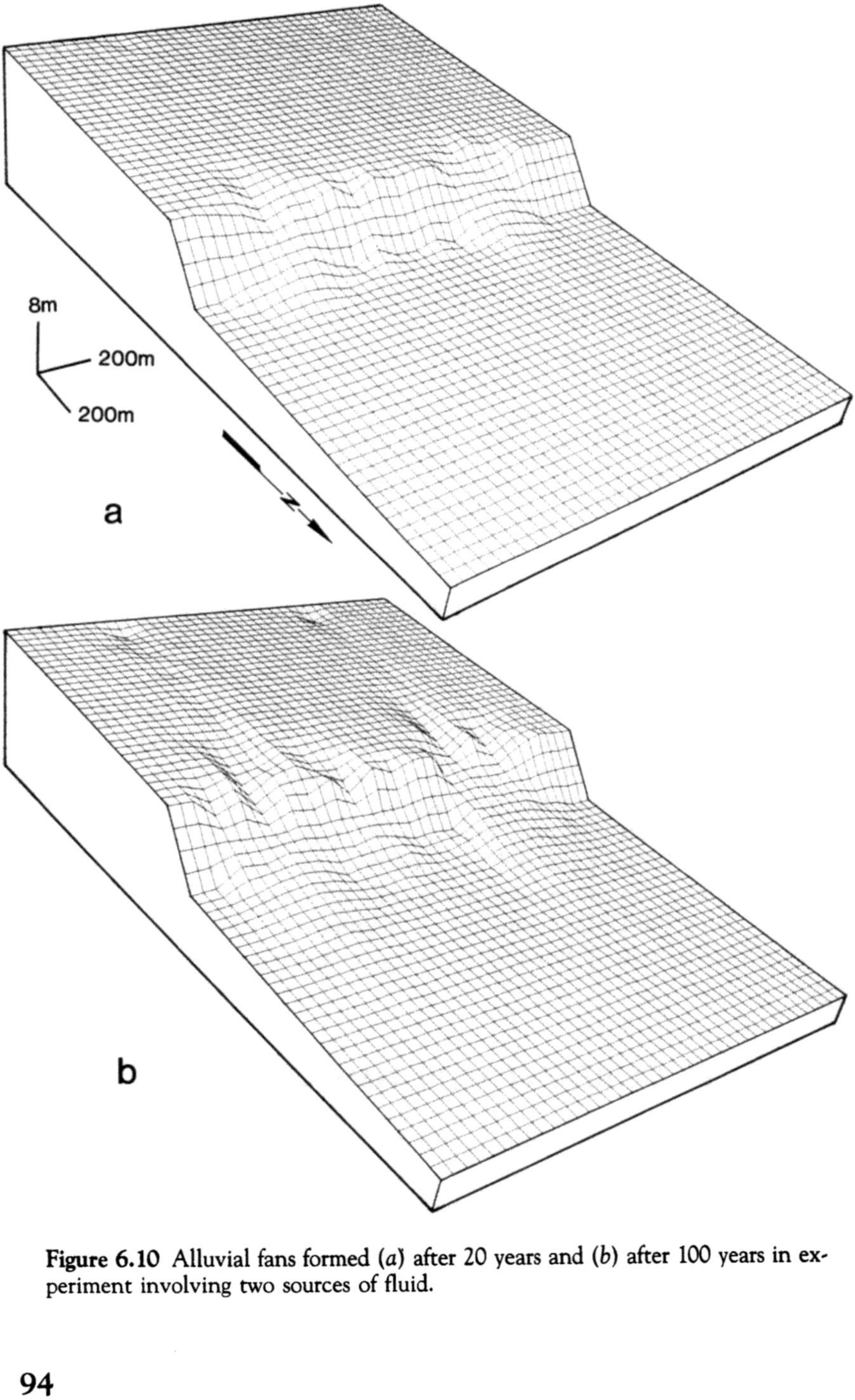

Figure 6.10 Alluvial fans formed (*a*) after 20 years and (*b*) after 100 years in experiment involving two sources of fluid.

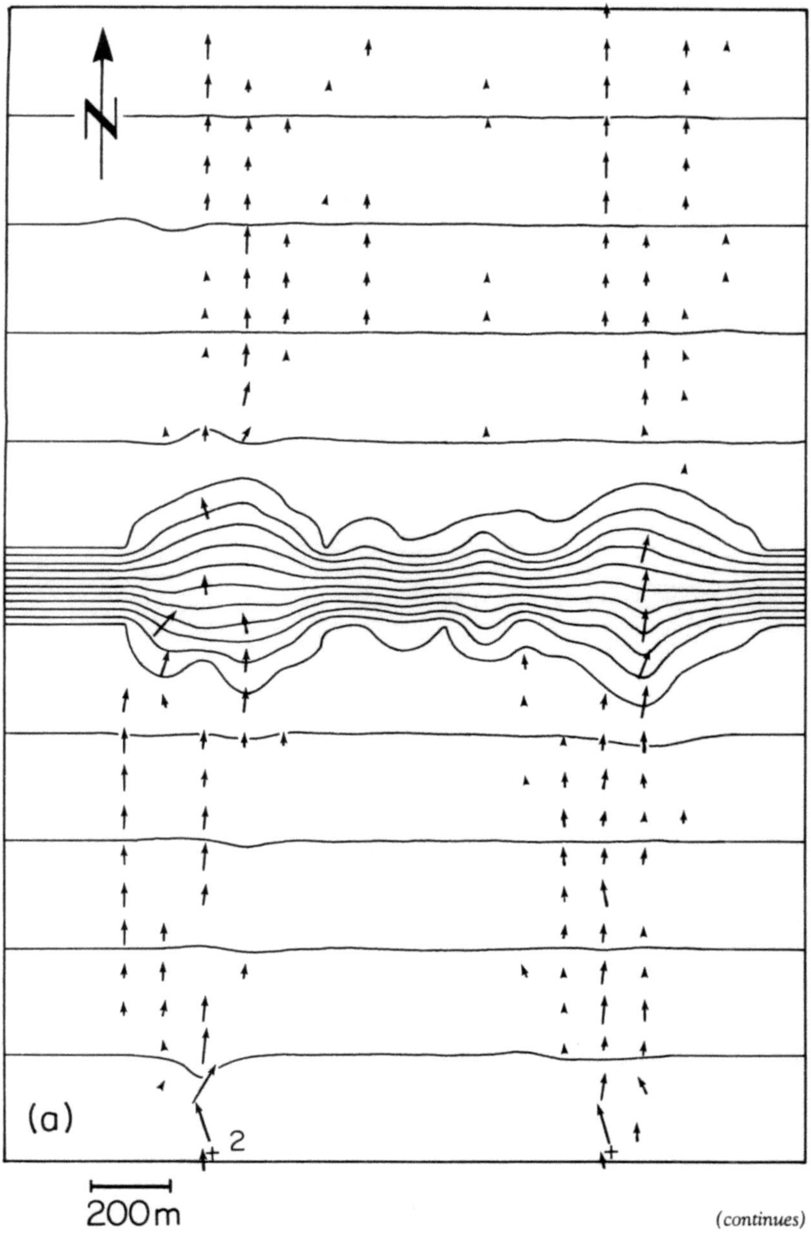

(continues)

Figure 6.11 Succession of maps showing topography and flow velocities in alluvial-fan experiment. (*a*) After 20 simulated years. (*b*) After 60 simulated years. (*c*) After 100 years. Contours in meters.

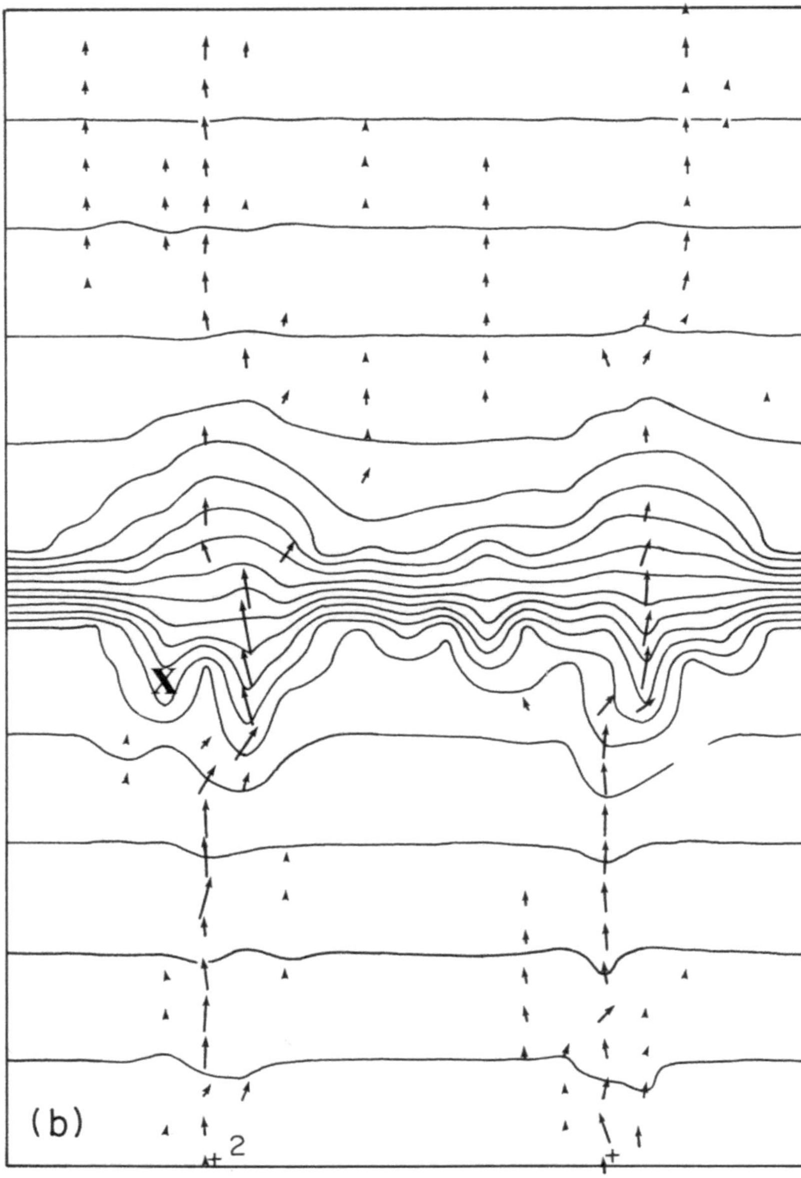

Figure 6.11 (continued)

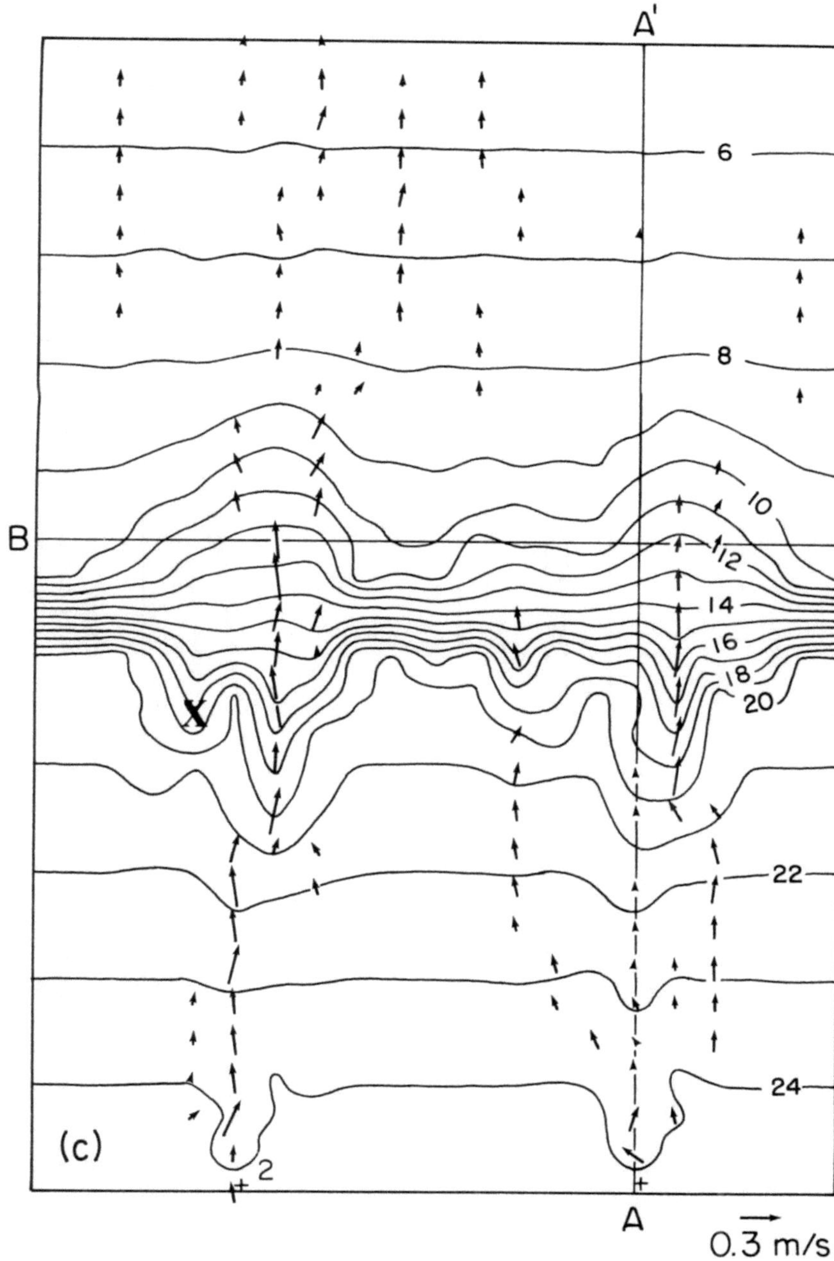

Figure 6.11 (continued)

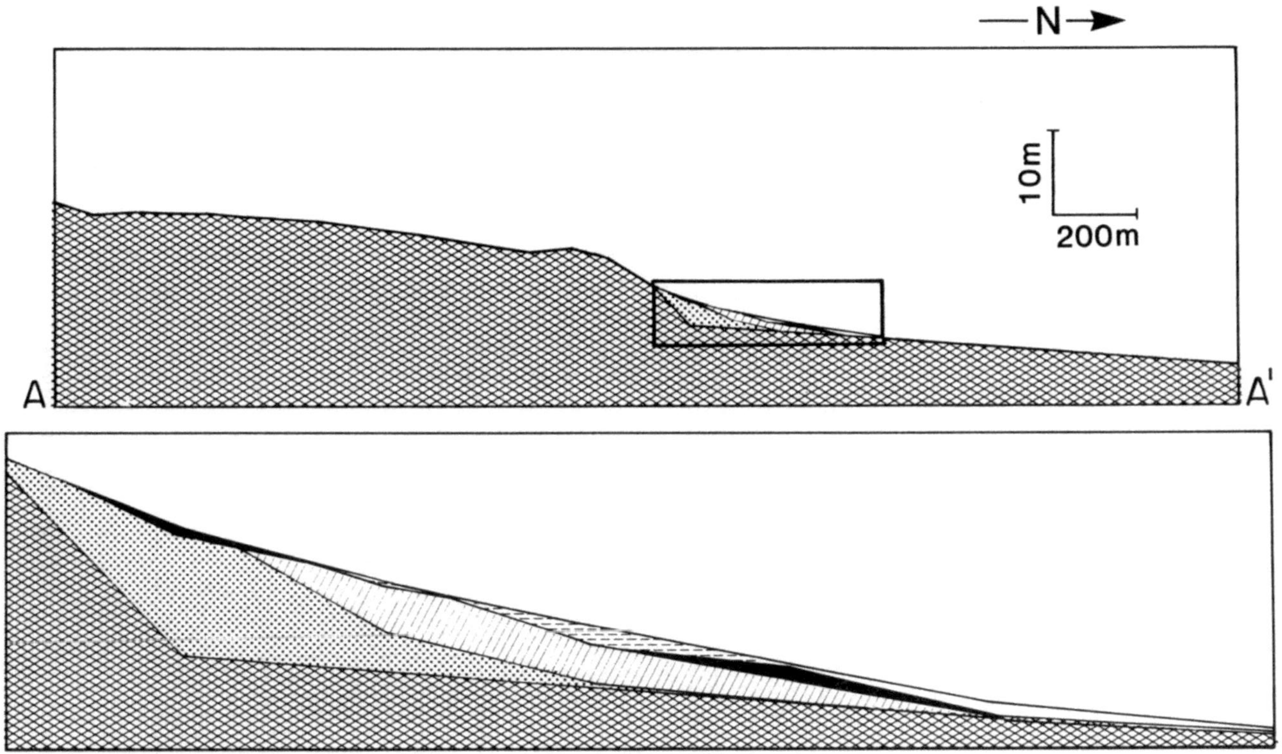

Figure 6.12 North-south section A-A′ in Figure 6.11 revealing irregular profile produced by erosion in southern part and smooth profile formed by deposition in northern part. Original bedrock is shown by diagonally ruled pattern. Other patterns show deposits formed within timelines drawn every 20 years (deposits are of uniform composition). (*a*) Entire profile. (*b*) Enlargement of part of profile.

The turbidity flow is more irregular than in the earlier, flow-only experiment. The irregularity is produced by interactions between the flow and the submerged topography, the flow both eroding and depositing. The flow is affected by changes in topography, but the flow also changes the topography, both being highly interdependent. The flow slows at the bottom of the steep slope, where deposition has formed a fan (Fig. 6.13*c*–*f*).

A second experiment involving 20 successive flows is shown in Figures 6.14 (pages 102–104) and 6.15 (page 105). Each successive flow was released when the preceding flow had been completed. Erosion has produced a deep canyon, and the material eroded has been redeposited as a fan. Figures 6.14*a* through *f* show changes at intervals separated by four intervening flows, and Figure 6.15 shows the final topographic form.

SEDSIM2'S OVERALL PERFORMANCE

Restriction of flow to two horizontal dimensions affects SEDSIM2's ability to simulate processes where vertical variations in velocity are important. For example, SEDSIM2 cannot simulate sedimentary features whose vertical dimensions are significantly smaller than flow depth, such as subaqueous ripples. However, SEDSIM2 can simulate the development of underwater features, such as dunes, caused by large-scale changes in flow velocity and depth. Although SEDSIM2 does not reproduce the helical flow in channel bends, SEDSIM2 nevertheless performs well in simulating erosion and deposition in channel bends, and in simulating the lateral shifting of channels. Part of SEDSIM2's effectiveness lies in the fact that SEDSIM2 deals with variations in shear stress that occur across channels, which in turn affect erosion and deposition. SEDSIM2 should be further calibrated so that its rates of deposition and erosion are comparable to those in nature.

SEDSIM2's performance might be improved by changing its sediment transport formula. But because transport mechanisms are not thoroughly understood, it is difficult to devise a formula that is flexible enough to represent sediment transport over a wide range of hydraulic conditions. For example, use of formulas that represent bed load and suspended load separately might increase the accuracy of SEDSIM2's sediment transport rate calculations, but it would significantly increase the computing time, because the computer program must evaluate the transport formula for each fluid element at each time increment.

In general, SEDSIM2 adequately reproduces major sedimentary processes in channels, alluvial fans, and turbidity currents. In most natural environments, however, sediment of various particle sizes is present, and thus it is necessary to extend SEDSIM2 to accommodate several particle sizes, a task described next.

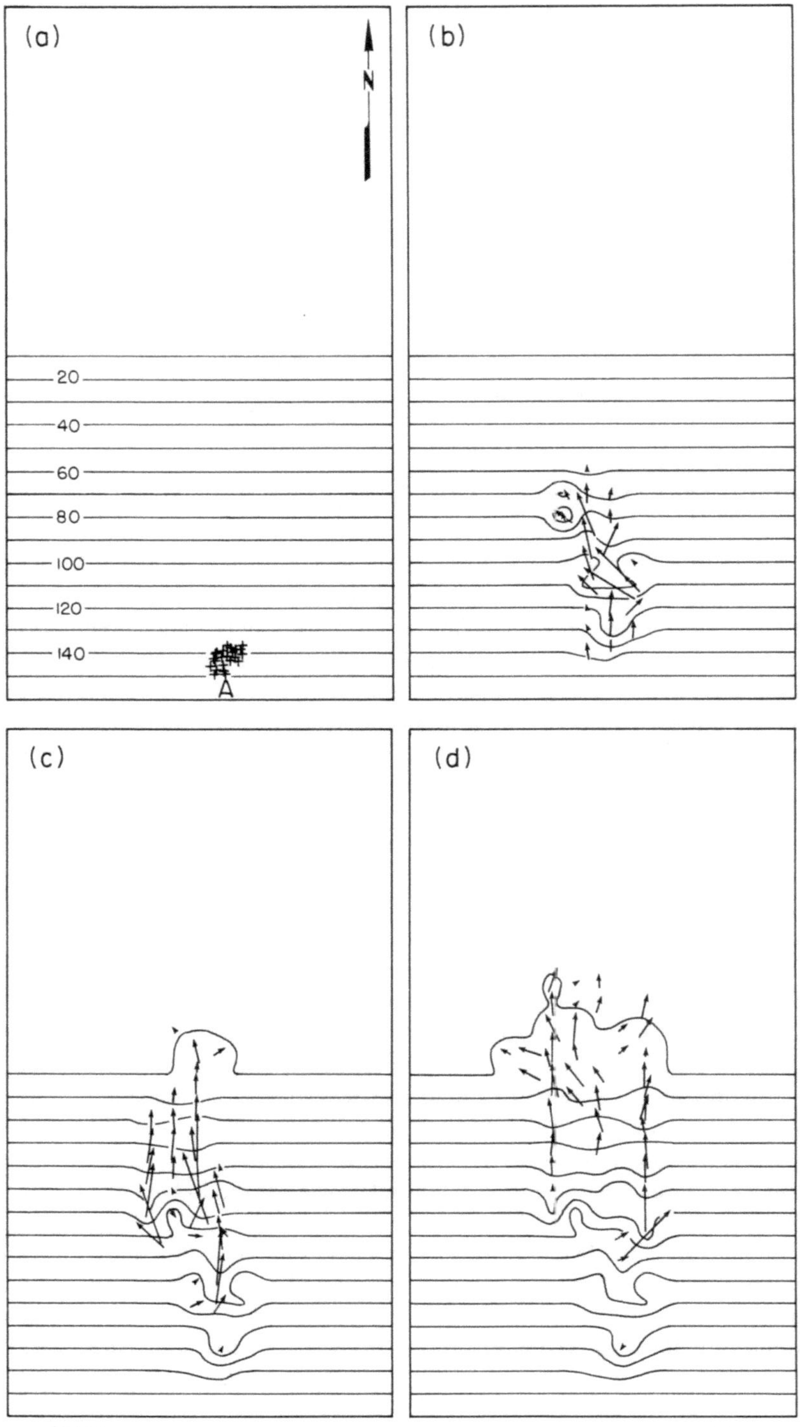
(a)
N
20
40
60
80
100
120
140
A
(b)
(c)
(d)
400m
1· m/s

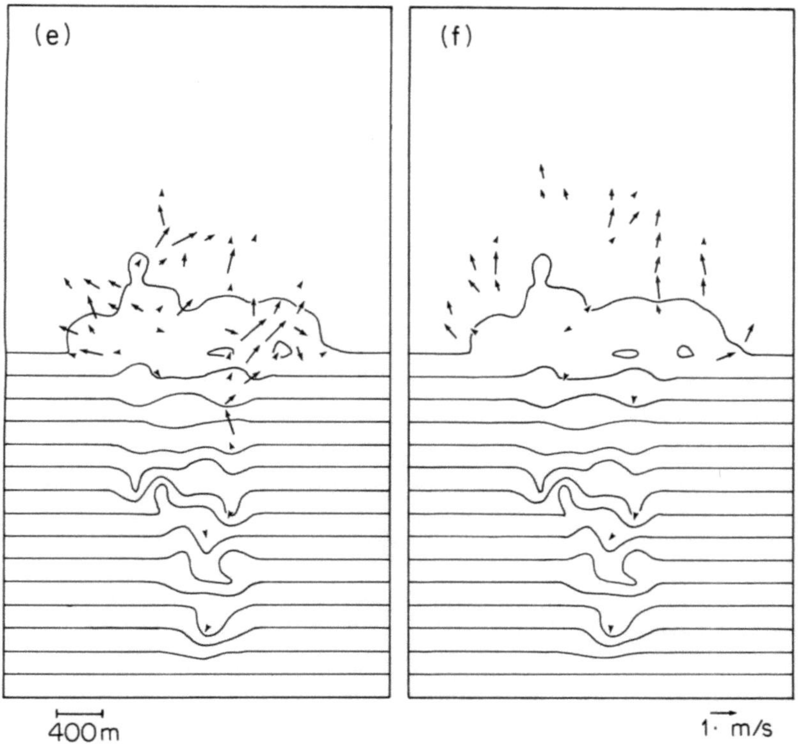

Figure 6.13 Contour maps and velocity vector plots (*a*) through (*f*) at intervals of one simulated minute showing progress of turbidity current in loose fine sand. Flow spreads out on gentle slope where depositional fan has formed. Erosion and deposition during flow cause flow to be much less regular than flow shown in Figure 4.12. Contours in meters above datum at north end of area. 'A' in (a) denotes location of source of turbidity current.

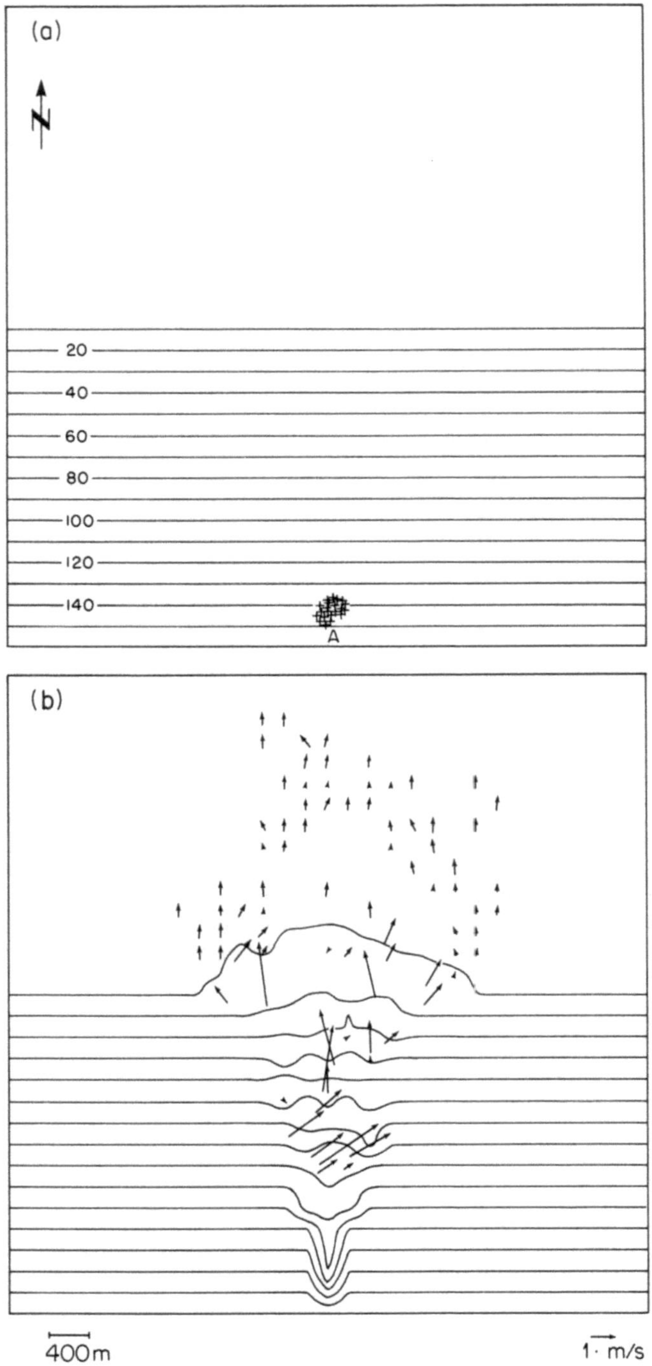

Figure 6.14 Contour maps and velocity vector plots (*a*) through (*f*) showing successive effects of 20 turbidity currents in loose fine sand. Each map is separated by interval of four turbidity flows. Contours in meters above datum at north end of area. 'A' in (a) denotes location of sources of turbidity currents.

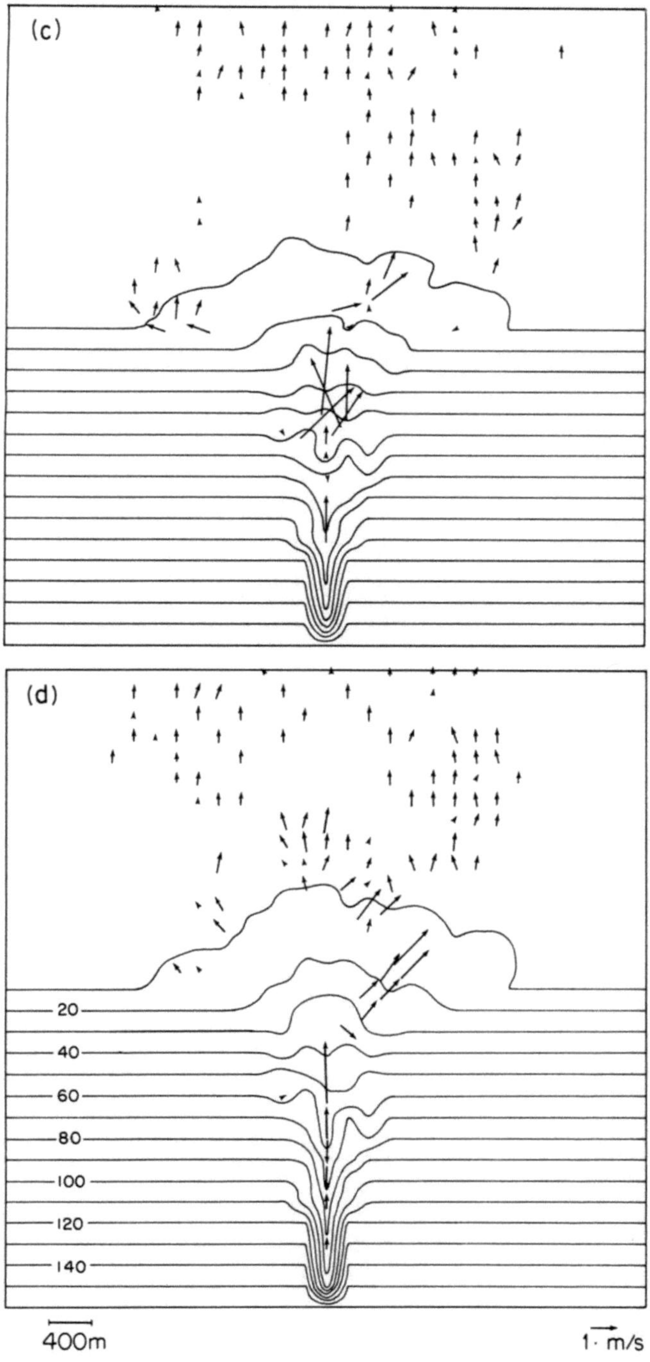

Figure 6.14 (continued)

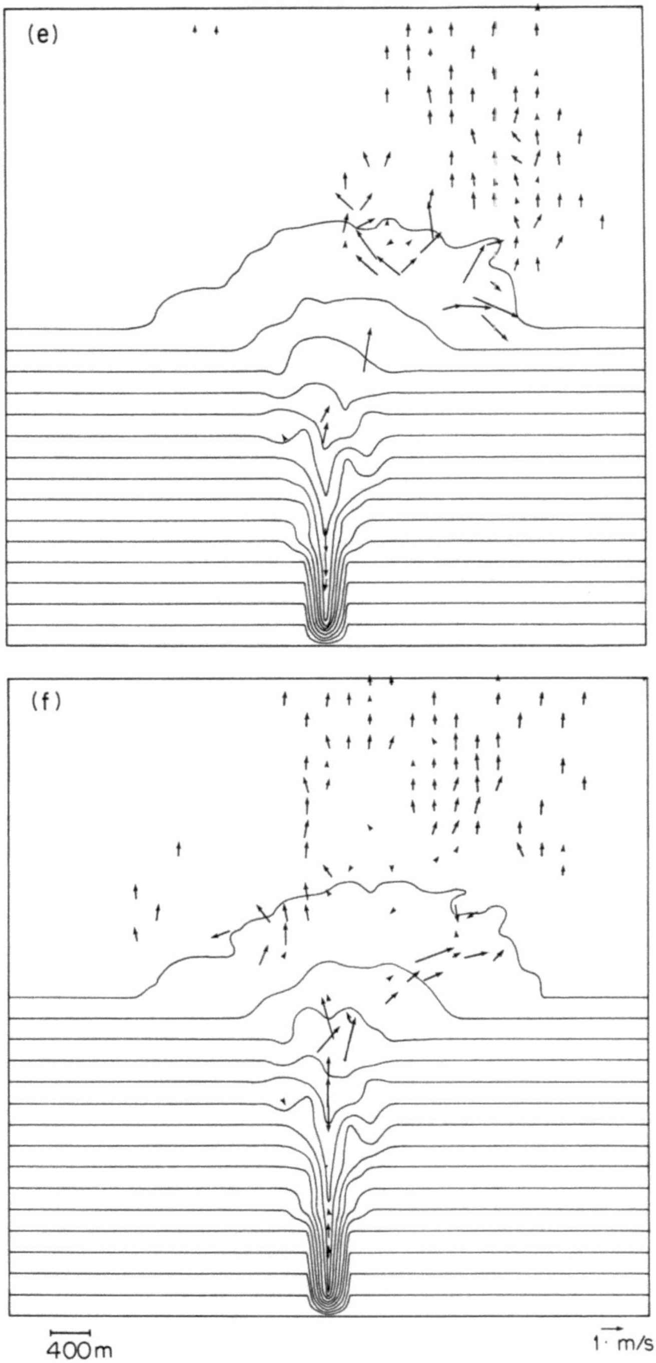

Figure 6.14 (continued)

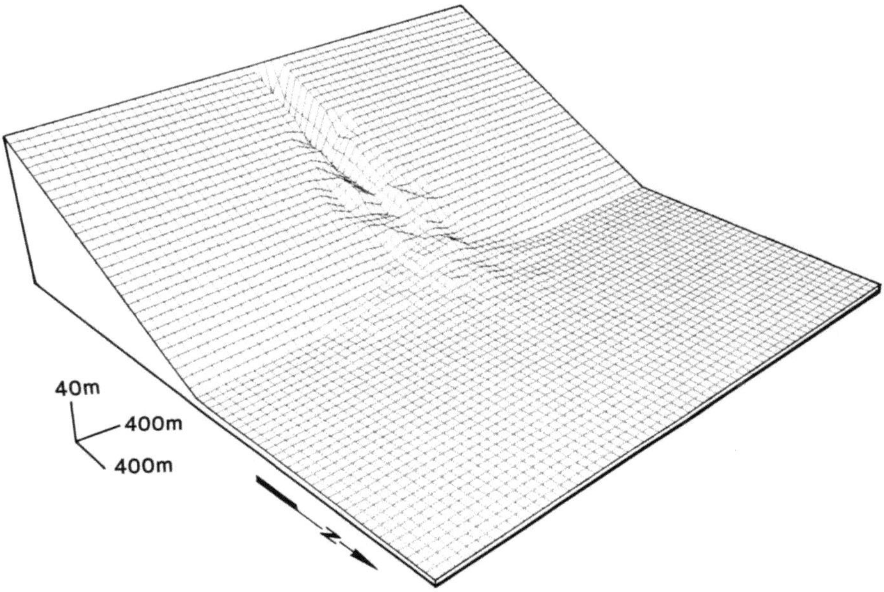

Figure 6.15 Canyon and fan at conclusion of experiment after 20 turbidity flows (corresponds to Figure 6.14*f*).

SUMMARY

We extended the program that simulates flow by incorporating the equations that govern sediment transport developed in Chapter 5. Because of the length of the simulation experiments, we devised algorithms that reduced the arithmetic performed by SEDSIM2. These procedures proved effective in experiments involving hundreds of years of simulated time, but they must be used cautiously to avoid unrealistic depositional or erosional effects, particularly near sources of fluid and sediment. The experiments that simulate sedimentary processes in a bending channel, alluvial fans, and turbidity currents, are more or less satisfactory in spite of SEDSIM2's inability to simulate flow in three full dimensions.

CHAPTER 7

SEDSIM3's Representation of Multiple Sediments

For realism we need a flow system capable of transporting sediment grains of various sizes. SEDSIM3 represents an extension of SEDSIM2, and it permits up to four different grain sizes to be transported, providing a more versatile simulation model that is applicable to a larger range of geological problems. However, in dealing with multiple grain sizes, we encounter two major challenges:

1. If different sizes of grains are carried in a flow, the flow's capacity to transport each size is affected by the presence of the other particle sizes. Furthermore, the interactions between different sediment sizes in the flow are not fully understood, creating a challenge in representing their behavior in a simulation model.
2. A multiple-sediment simulation model must represent variations in sediment that has been deposited, requiring that accurate accounts be kept. Although procedures that represent three-dimensional bodies of sediment are simple in principle, they involve extensive arithmetic and can consume a large proportion of the computer memory required for experiments.

GRAPHIC DISPLAY OF DEPOSITED SEDIMENT IN THREE DIMENSIONS

The mathematical procedures for displaying deposits containing multiple types of sediment are presented here. These procedures are valid for any number of types of sediment, although for simplicity SEDSIM3 accommodates a maximum of four types. The limit of four is a compromise between the real world, where a continuum of particle sizes may exist, and computing requirements, which can become very large if there are many sediment types. Furthermore, an advantage of a limit of four is that a mixture of any proportion of four components can be represented unambiguously with a single color, greatly facilitating graphic representation.

SEDSIM3's function that represents the three-dimensional composition of deposited sediment is

$$\kappa(x, y, z) = K_s \tag{7.1}$$

where

κ = function assigning a sediment type to every point underneath the surface
K_s = sediment type
(x, y, z) = point underneath the surface ($z \leq Z$)

Each sediment type is represented by an integer because the number of sediment types is finite, rather than a continuous gradation. Only a single type is represented at any point within the three-dimensional volume forming the "subsurface." Thus, at a single point, SEDSIM3 does not represent sediment mixtures, because mixtures can be defined only when a volume of sediment consisting of an aggregate of points is considered. The representation of a single sediment type at each point is partly justified because only a single grain of sediment can be present at any point in an actual deposit.

CONTINUITY EQUATION FOR MULTIPLE TYPES OF SEDIMENT

The continuity equation for multiple types of sediment is an extension of the continuity equation for sediment of a single grain size (Eq. 5.1). The extended equation specifies that the volume of material represented by net changes in topographic elevation is equal to the net change in the load of all sediment types:

$$(H - Z) \sum \frac{Dl_{K_s}}{Dt} = -\frac{\partial Z}{\partial t} \tag{7.2}$$

where

H = free surface elevation with respect to sea level
Z = topographic elevation with respect to sea level

Σ = summation over the number of sediment types
l_{K_s} = volumetric sediment concentration of type K_s
K_s = sediment type
t = time

When erosion occurs, only that material immediately adjacent to the water-sediment interface is eroded. When deposition occurs, the newly deposited sediment becomes the sediment type immediately below the water-sediment interface. These relationships can be expressed as

$$\frac{Dl_{K_s}}{Dt} \neq 0 \Rightarrow \kappa(x, y, Z) = K_s \tag{7.3}$$

FORMULAS FOR TRANSPORTING SEDIMENT OF DIFFERENT GRAIN SIZES

The general form of the equation for transporting multiple types of sediment is

$$(H - Z)\frac{Dl_{K_s}}{Dt} = f(Q, \nabla H, \nabla Z, L, F, K_s, \kappa(Z)) \tag{7.4}$$

where

L = vector defining sediment concentration of each type: $(l_1, l_2, \ldots l_{N_s})$
F = matrix of coefficients defining sediment types:

$$\begin{bmatrix} f_{1,1} & f_{1,2} & \cdots & f_{1,N_f} \\ f_{2,1} & f_{2,2} & \cdots & f_{2,N_f} \\ \cdots & & & \\ f_{N_s,1} & f_{N_s,2} & \cdots & f_{N_s,N_f} \end{bmatrix}$$

N_f = number of coefficients that define each sediment type
N_s = number of sediment types

As in the single-sediment version, the general form of the sediment transport equation does not state how much sediment is transported until the function f is defined. To define f, we can adapt existing formulas for quantifying sediment transport for mixtures of sediment. Two useful formulas are provided by Kalinske (1947) and by Laursen (1958). Kalinske's formula is

$$Q_s = \frac{A_2}{A_1} r v_* d \tag{7.5}$$

where

Q_s = sediment transport rate
A_1 = ratio of grain area to grain diameter squared
A_2 = ratio of grain volume to grain diameter cubed
r = relative excess shear stress = $(\tau_0 - \tau_c)/\tau_c$
τ_0 = bottom shear stress
τ_c = critical shear stress
v_* = shear velocity
d = sediment particle diameter

When Kalinske's formula is applied to a mixture of several particle-sized fractions, the transport rate for each fraction is given by Equation 7.5, and then the result is weighted by a "coverage factor." The coverage factor represents the proportion of the water-sediment interface that is covered by sediment grains of a specific size, and is proportional to p_i/d_i, where p_i is the proportion by weight of sediment fraction i, and d_i is the particle diameter of that fraction.

Laursen's (1958) formula is also useful:

$$\bar{c} = \sum_i p_i \left(\frac{d_i}{h}\right)^{7/6} \left(\frac{\tau_0'}{\tau_c} - 1\right) f\left(\frac{v_*}{W}\right) \tag{7.6}$$

where

$\bar{c}$ = mean concentration of suspended load in percentage by weight
p_i = proportion by weight of fraction of size d_i
τ_0' = shear stress due to grain resistance alone
= $v^2 d_i^{1/3}/30\, h^{1/3}$
W = particle fall velocity
f = function determined experimentally

Kalinske's and Laursen's formulas are simple to apply because the interactions between sediment types are determined by the coverage factor or by the proportion by weight of each sediment type. SEDSIM3's formula described below is similar, but to maintain compatibility with the single-sediment formula used in SEDSIM2, SEDSIM3's formula assumes that the main parameter determining total transport rate is power dissipated by bottom friction per unit volume of fluid (P), rather than relative excess-shear stress (r).

Procedures provided by other authors for calculating the transport rate of components of a sediment mixture are more accurate, but require more computation than Kalinske's and Laursen's formulas. The sediment transport formula used by SEDSIM3 is evaluated during every time increment for every fluid element, and therefore it must be simple to minimize computing effort. Appropriate representation of sediment transport is vital, however, because whether erosion or deposition occurs under given hydraulic conditions depends on total sediment load and on composition of the load.

SEDSIM3'S FORMULA FOR MULTIPLE-SEDIMENT TRANSPORT

The formula that SEDSIM3 uses for transporting multiple types of sediment is similar to the formula for a single sediment type represented by Equation 5.8. Both formulas share the assumption that erosion and deposition depend on the balance between the flow's transport capacity, and the "effective sediment concentration," defined for a single sediment type in Equation 5.8. For the multiple-sediment situation, we must compare the effects of different sediment mixtures on a flow's capability to transport them. This requires that we define the effective sediment concentration for specific sediment mixtures, Λ_{em}. In SEDSIM3 we assume that Λ_{em} is the sum of the values of Λ_e for each component, as defined by Equation 5.8. Therefore

$$\Lambda_{em} = \sum_{K_s} \frac{l_{K_s}}{f_{1,K_s}} \tag{7.7}$$

where

Λ_{em} = effective sediment concentration of mixture
l_{K_s} = sediment concentration of each type
f_{1,K_s} = transportability of each sediment type (as defined in Chapter 5)

As in the single-sediment model, SEDSIM3's transport capacity is defined by Equation 5.10. However, the equivalent of the single-sediment transport equation (Eq. 5.8) in SEDSIM3 is expressed by Equations 7.8 and 7.9. Let

$$R = (\Lambda - \Lambda_{em})f_{2,K_s} \tag{7.8}$$

where f_{2,K_s} = erosion-deposition rate coefficient for sediment type K_s. Then for every sediment type K_s;

$$(H - Z)\frac{Dl_{K_s}}{Dt} = \begin{cases} R & \text{if } R > 0 \text{ and } \tau_0 \geqslant f_{3,K_s} \text{ and } \kappa(x, y, Z) = K_s \\ & \text{or } R < 0 \text{ and } K_s = 1 \text{ or } l_{K_s-1} = 0 \\ 0 & \text{otherwise} \end{cases} \tag{7.9}$$

where

K_s = sediment type
$K_s - 1$ = next coarser sediment type

Equation 7.9 specifies that when the effective load is less than transport capacity, material is eroded if the threshold conditions for moving the exposed sediment are equalled or exceeded. As in single-sediment SEDSIM2, the transport equation is modified to prohibit erosion in hypopycnal flow. When effective load exceeds transport capacity, deposition occurs and coarser sediments (i.e., sediment particles with higher fall velocities) are deposited first. Erosion and deposition rates are proportional to the difference between effective load and transport capacity.

Equation 7.9, which regulates erosion, transport, and deposition, has two advantages: (1) It is compatible with the single-sediment transport equation (Eq. 5.8). If only one sediment type is used in Equation 7.9, or if all sediment types are the same, Equations 5.8 and 7.9 are identical. (2) If the effective sediment concentration of a sediment mixture is defined as the sum of the effective concentrations of its components (Eq. 7.7), we need not calibrate other sediment parameters for the multisediment model, because they can be obtained from the single-sediment model.

SEDSIM3'S INITIAL CONDITIONS AND BOUNDARY CONDITIONS

Initial conditions for operating SEDSIM3 are similar to those for SEDSIM2 (Eq. 5.14), except that sediment concentrations for each type of sediment must be provided for SEDSIM3 as follows:

$$l_{K_s}(x, y, t_0) = l_{K_s,0}(x, y) \qquad 1 \leq K_s \leq N_s \tag{7.10}$$

Also, the type of sediment existing in every cell must be prescribed at the start of a simulation experiment:

$$\kappa(x, y, z, t_0) = \kappa_0(x, y, z) \tag{7.11}$$

If sources of fluid and sediment are to be supplied as boundary conditions, the concentration of each sediment type must be specified at each source:

$$l_{K_s}(x_{s_i}, y_{s_i}, t) = l_{K_s, s_i}(t) \tag{7.12}$$

THE BASEMENT

In many experiments we assume that sedimentary deposits rest upon homogeneous material that continues indefinitely downward, thus constituting the "basement." The basement can be eroded, however, thus yielding one or more types of sediment. In SEDSIM3, the basement is not necessarily firm and resistant, because it can be composed of erodible material. Thus, basement simply denotes uniform material that extends downward without limit.

Basement thus defined has advantages because it is necessary to retain only a single item of information about its composition. The only other information pertains to the shape and elevation of its upper surface. Because we must specify κ everywhere in the subsurface, we can arbitrarily assume it to be 0 for cells containing basement. Conveniently, the computing procedure that evaluates κ can identify material of type 0 much faster than the other types of deposits simply by noticing that the cell to be evaluated is below the basement's upper surface. Such a procedure reduces both storage requirements and computing time.

If basement is to be included, Equation 7.9 must be expanded to deal with the possibility that the basement may be eroded:

Let

$$R_0 = (\Lambda - \Lambda_{em})f_{2,0} \tag{7.13}$$

$$(H - Z)\frac{Dl_{K_s}}{Dt} = \begin{cases} R_0 f_{4,K_s} & \text{if } R_0 > 0 \text{ and } \tau_0 \geqslant f_{3,0} \text{ and } \kappa(x, y, Z) = 0 \\ \text{defined by Equation 7.9 otherwise} \end{cases} \tag{7.14}$$

where

$f_{2,0}$ = basement erosion coefficient
$f_{3,0}$ = threshold shear stress for basement erosion
f_{4,K_s} = proportion of sediment type K_s that the basement yields when eroded, $1 \leqslant K_s \leqslant N_s$

SEDSIM3'S NUMERICAL METHOD FOR REPRESENTING MULTIPLE SEDIMENTS

SEDSIM3's representation of fluid flow is identical to that of SEDSIM2, even though multiple types of sediment are represented. Equations 3.1 through 3.12 which deal with flow and which were utilized in flow-only SEDSIM1, and also in single-sediment SEDSIM2, remain valid for multiple sediments. However, new equations are needed to represent the sediments that have been deposited as three-dimensional bodies whose properties vary from cell to cell.

The representation involves a gridwork of cells forming the "subsurface," in which each cell is endowed with a specific single type of sediment. A two-dimensional array represents the elevation of the upper surface of the basement, and another two-dimensional array represents the elevation of the upper surface of the body of sediment (the topographic grid).

The volume between these two two-dimensional arrays contains the deposited sediment and is represented by a three-dimensional array (the sediment body array). The function κ assigns a specific type of material to each cell in the sediment body array, each type of sediment being represented in each cell by a positive integer, whereas the basement is assigned a value of 0. Cells in the sediment body array are stacked in prisms between the top of the basement and the topographic surface, the prisms being square in horizontal section (Fig. 7.1). The cells contained in the sediment body array have the same horizontal dimensions as cells that define the top of the basement and as cells that define the topographic surface. However, the sediment body cells are displaced by half a cell's width in both the x and y directions. Therefore, the corners of topographic cells, where hydraulic parameters are calculated, are directly underneath the centers of sediment cells, where erosion and deposition take place. The vertical dimension of each three-dimensional cell that contains sediment is fixed, except that the uppermost cell in a prism may be thinner to accommodate the specific topographic elevation defined by the corresponding cell in the topographic grid.

Sediment Continuity

When multiple sediment types are represented, we must specify the concentration for each sediment type K_s, for each fluid element k, at each time t, which can be written as $l_{k,K_s,t}$. The continuity equation that defines the conservation of material then can be approximated in finite-difference form as

$$\frac{Z_{i,j,t+1} - Z_{i,j,t}}{\Delta t} = -h_{i,j,t} \sum_{k} \sum_{K_s} \frac{l_{k,K_s,t+1} - l_{k,K_s,t}}{\Delta t} \qquad (7.15)$$

Note that variables in Equations 7.15 through 7.21 are defined as they were for the single-sediment model unless otherwise stated.

Sediment Transport

The sediment transport equations (Eqs. 7.9 and 7.14) can be approximated with algebraic expressions, employing the following definitions:

1. $D(K_s, i, j, z)$ is the thickness of sediment type K_s contained above level z and underneath the topographic surface at grid point (i, j).

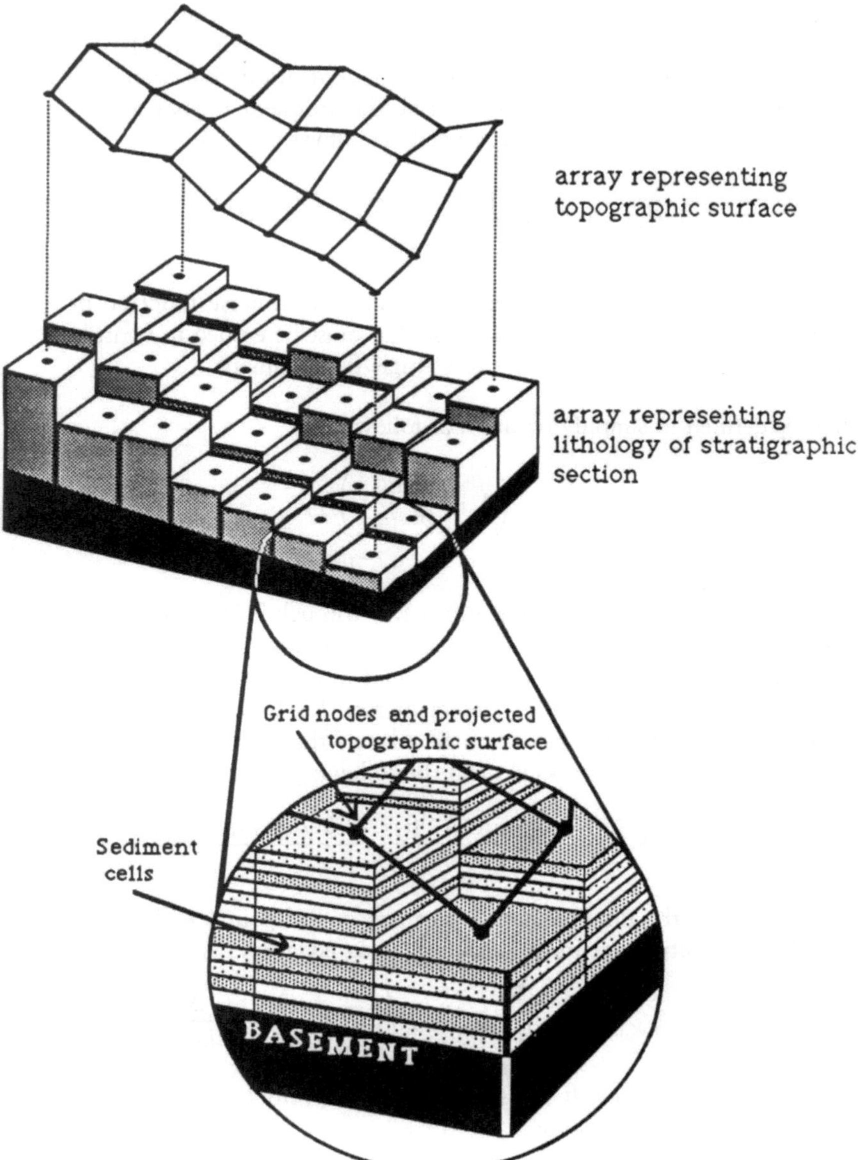

Figure 7.1 Diagram showing how SEDCYC3 records distribution of deposited sediment in three dimensions.

2. $D_\tau(\tau, i, j)$ is the depth below the surface of the highest point whose threshold for erosion (f_{3,K_s}) is greater than τ at grid point (i, j).
3. $\Lambda_c(i, j, z)$ is the effective sediment concentration that would be contributed to a single fluid element if all the sediment (or basement) above level z at grid point (i, j) was eroded, as follows:

$$\Lambda_c(i,j,z) = \sum_{K_s} \left(\frac{D(K_s, i, j, z)}{h_e f_{1,K_s}} + \frac{D(0, i, j, z)}{h_e f_{1,K_s}} \right) \tag{7.16}$$

where h_e is the fluid element depth (V_e/A). The inverse function of Λ_c is $\Lambda_c^{-1}(i, j, l)$, and represents the thickness that must be eroded to contribute an effective sediment concentration l to a single fluid element.

4. C is defined to simplify notation as follows:

$$C = \left(\Lambda - \sum_{K_s} \frac{l_{k,K_s,t}}{f_{1,K_s}} \right) f_{2,K_s} \tag{7.17}$$

where $\Lambda = |\mathbf{Q}_{i,j}|^3 c_1 c_t$.

Now it is possible to define Equations 7.18 through 7.21, which provide the expressions used by SEDSIM3 to approximate the sediment transport equations. Equations 7.18 through 7.21 are applied at every time increment t to every fluid element k and to every sediment type K_s. If $C > 0$ and if the fluid element is not near a grid point on the edge of the grid (i.e., if $1 < i < M$, $1 < j < N$), then erosion occurs and the following equation applies:

$$l_{k,K_s,t+1} = l_{k,K_s,t} + \frac{D(K_s, i, j, \min(\Lambda_c^{-1}(i, j, C\,\Delta t/h_e), D_\tau(\tau_{i,j}, i, j)))}{h_e} \tag{7.18}$$

If $C < 0$ and if the fluid element is not near a grid point on the edge of the grid, then deposition occurs, and the following equation applies:

$$l_{k,K_s,t+1} = l_{k,K_s,t} - \min\left[l_{k,K_s,t}, \max\left(\left(\sum_{K=1}^{K_s} l_{k,K,t} \right) - \frac{C\,\Delta t}{h_e}, 0 \right) \right] \tag{7.19}$$

When deposition occurs, it is also necessary to specify the sediment type that fills each new cell, as follows:

$$\kappa_{i,j,n_z} = K_s \qquad \text{if} \begin{cases} K_s = 1 \text{ or } z_b + n_z z_c > h_e \sum_{K=1}^{K_s - 1} l_{k,K,t} - C\,\Delta t \\ \text{and} \\ z_b + n_z z_c > h_e \sum_{K=1}^{K_s} l_{k,K,t} - C\,\Delta t \end{cases} \tag{7.20}$$

where

κ_{i,j,n_z} = sediment type at cell (i, j, n_z)
n_z = sediment cell number above basement
z_b = elevation of basement top
z_c = vertical dimension of one sediment cell

If C = 0 (or if the grid point nearest element k is on the edge of the grid), then no change occurs:

$$l_{k,K_s,t+1} = l_{k,K_s,t} \tag{7.21}$$

PROGRAM SEDCYC3

Following the nomenclature of previous versions, the program that performs the simulation in the multisediment model has been called SEDCYC3. SEDCYC3 transports up to four types of sediment and represents them in a three-dimensional body of deposited sediment that lies beneath the topographic surface and above the basement. Basement is represented by a body of uniform composition whose shape is completely specified by elevation of the upper surface. When basement material is eroded, the elevation of the top of the basement is reduced, but the elevation of the basement top cannot increase, because "additions" to basement do not occur in the sense that "subtractions" by erosion occur (although future extensions of SEDSIM could incorporate structural deformation, thereby permitting the top of the basement to be elevated or lowered when structural deformation occurs).

The three-dimensional sediment body grid extends between the top of the basement and either the land surface or the water-sediment interface (Fig. 7.1). However, there is an optimal provision to constrain the grid so that it contains only sediment that is actively eroded or deposited during a simulation experiment, permitting the vertical dimensions of the sediment grid to be reduced.

Storage of deposited sediment economizes memory requirements by representing each of the four sediment types with only two bits of information, instead of using a 32-bit integer. This procedure is known as "packing" and allows the contents of 16 sediment cells to be represented in the same memory space that would be required for a single sediment cell if 32-bit integers were used. However, packing causes execution time to increase.

INPUT TO SEDCYC3

SEDCYC3 employs a longer input data file than SEDCYC2. An example is presented in Table 7.1, which can be contrasted with the lesser requirements for SEDCYC2 in Table 6.1. SEDCYC3's input and output files contain grids for the top of the basement and for the topographic surface, as well as for the three-dimensional sediment body grid. As with SEDCYC1 and SEDCYC2, SEDCYC3 produces two output files. One consists of the final state of the sys-

Table 7.1 Example Input Data File for SEDCYC3

```
TITLE: MULTI-SEDIMENT EROSION AND DEPOSITION ON INCLINE

RUN PARAMETERS:
START TIM=0.00E+00 Y
END TIM  =0.10E+00 Y
DISPL INT=0.10E+00 Y
TIM INCR1=0.50E+01 S
TIM FACT1=0.10E+04
TIM INCR2=0.10E+00 Y
TIM IDLE =0.00E+00 Y
TIM FACT2=0.10E+01

GENERAL PHYSICAL PARAMETERS:
FLOW DENS=    1000. KG/M3
SEA DENS =    1027. KG/M3
LAT FRICT=     100. NS/M2
ROUGHNESS=  0.0400
EL VOLUME=0.10E+05 M3
CURR(X,Y)=0.00E+000.00E+00M/S

SEDIMENT PARAMETERS:
                 S1       S2       S3       S4      BAS
DIAMETER = .50E-01  .10E-02  .80E-02  .10E-04  -----   M
DENSITY  =   2700.    2700.    2700.    2700.  -----   KG/M3
BAS DECAY=   0.200    0.300    0.300    0.200  -----
COHESION = .60E+00  .50E+00  .40E+00  .50E+00  .50E+00

SOURCES:
INTERVAL = 1.0 E+10Y    0.50E-04 Y
N SOURCES=          1
  X(M)      Y(M)     XV(M/S) YV(M/S)   S1(M3)   S2(M3)   S3(M3)   S4(M3)
  1000.     2675.     -0.5     -1.5   .80E+040.80E+040.80E+040.80E+04

TOPOGRAPHY:
GRID SIDE=    100.0 M
NROWS    =        31
NCOLS    =        31

GRID NODES ELEVATION (SURFACE) (M)
    50.0     50.0     50.0     50.0     50.0
    45.0     45.0     45.0     45.0     45.0
    40.0     40.0     40.0     40.0     40.0
    35.0     35.0     35.0     35.0     35.0
    30.0     30.0     30.0     30.0     30.0
```

(continued)

```
        (continued)

GRID NODES ELEVATION (BASEMENT) (M)
    51.0    51.0    51.0    51.0    51.0
    45.8    45.4    45.0    45.0    45.0
    40.0    40.0    40.0    40.0    40.0
    35.0    35.0    35.0    35.0    35.0
    30.0    30.0    30.0    30.0    30.0

LITHOLOGIC COLUMNS:
LCELL DEP=    0.200
  5 11111
  5 11112
  5 11123
  5 11123
  5 11234
  4 1224
  2 24
  0
  0
  0
  0
  0
  0
  0
  0
  0
  0
  0
  0
  0
  0
  0
  0
  0
  0

ELEMENT POSITIONS AND VELOCITIES:
N OF ELEM=        0
   X(M)      Y(M)     XV(M/S) YV(M/S)   S1(M3)   S2(M3)   S3(M3)   S4(M3)
```

tem, and it is written in the same format as the input file so that it can be used as an input for a succeeding experimental run if desired. The other contains information for graphic display as input to the graphics programs.

SEDCYC3'S REPRESENTATION OF EROSION, TRANSPORT, AND DEPOSITION

Algorithms that handle multiple types of sediment in SEDCYC3 are almost entirely contained in SEDCYC3's subroutine NEWELM3, which is considerably longer than corresponding NEWELM2 used by SEDCYC2. At each time increment and for every fluid element, NEWELM3 calculates fluid transport capacity and effective sediment concentration. NEWELM3 also identifies the sediment composition in each of the cells that lie directly beneath the water-sediment interface. Then NEWELM3 decides whether sediment is to be eroded or deposited, or if there is to be no change, and NEWELM3 selects one of the following actions:

1. If erosion is to occur, Equation 7.18 applies, and the elevation of the topographic grid is lowered and the corresponding volume of each sediment type that has been eroded is added to the flow (Eq. 7.15). Only the cells where erosion occurs transfer part or all of their contents to the flow. If basement is eroded, values in the basement grid are reduced.
2. If deposition occurs, then Equations 7.19 and 7.20 apply. Each sediment type carried by the flow is deposited until deposition removes the flow's excess sediment concentration (Eq. 7.19). The decreased concentration of transported sediment is balanced by increased topographic elevation (Eq. 7.15), with new sediment added to the three-dimensional sediment grid (Eq. 7.20).
3. If neither erosion or deposition occur, then no changes in sediment load or topographic elevation take place (Eq. 7.21).

GRAPHIC OUTPUT

SEDSHO3 is the graphics program for SEDCYC3. SEDSHO3 operates similarly to SEDSHO2 used with SEDCYC2, in that successions of maps, vertical sections, and block diagrams are produced. However, when SEDSHO3 provides a vertical section, two drawings are produced, one placed above the other. The upper drawing represents the ages of deposits, and the lower represents the types of sediment that have been deposited. The lower drawing is commonly termed a *facies* section.

The colors that represent types of sediment are specified as input. Although each cell in the sediment grid contains only a single "pure" type of sediment, mixtures of sediment are represented with colors that pertain to clusters of adjacent cells. The graphics program calculates the sediment proportions within each cluster (Fig. 7.2). The clusters extend vertically between consecutive time lines, but their horizontal expanse is specified by the user. Each time line represents either the present or a former topographic surface. The user specifies the intervals at which successive time lines are to be drawn.

The display showing sediment age consists of successions of intervals bounded by time lines, each interval having been given a contrasting color. Time lines can be destroyed by erosion and removed from the stratigraphic record. Where two consecutive time lines are preserved, they bound the stratigraphic interval that represents deposits formed during the time interval between the two time lines.

Sections that display sediment composition have the same general form as sections that display age, except that the colors in the layers vary laterally in

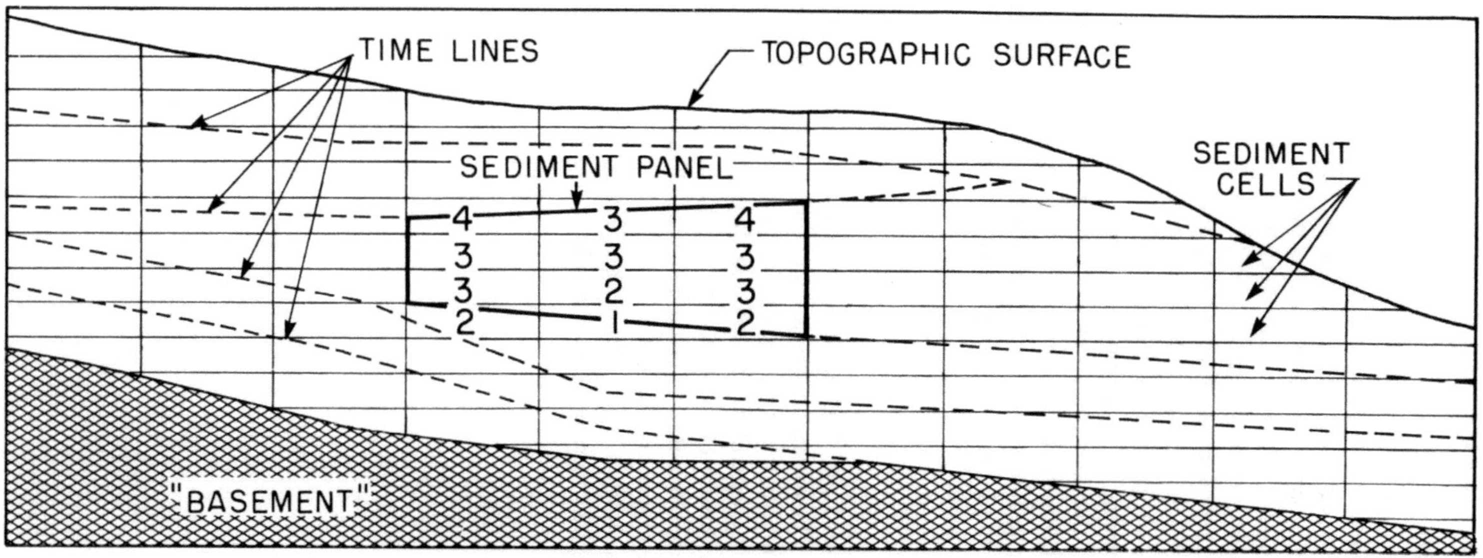

Figure 7.2 Vertical section showing how SEDSHO3 computes average sediment composition for group of adjacent sediment cells. Mixture can be then represented with single color utilizing color tetrahedron shown in Plate 1.

response to changes in sediment composition, thereby representing facies changes. Clusters of adjacent cells that contain sediment are represented by panels. The color assigned to a panel is the blending of types of sediment in cells spanned by the panel. In selecting the color, SEDSHO3 uses proportions of the three coarsest components and matches them to intensities of the three additive primary colors (red, green, and blue). The resulting color unambiguously represents a four-component mixture (Plate 1) because the fourth or finest component is represented either as white or black, depending on whether the background is black or white. The fourth component is necessarily black when "hard copies" are printed on white paper.

SEDSHO3 incorporates a mapping procedure that permits facies maps to be drawn representing the composition of sediment in a layer of constant thickness immediately beneath the topographic surface. The facies maps also employ a blending procedure similar to that used for sediment composition sections. Facies maps can be drawn at time increments specified by the user.

SUMMARY

We have extended the single-sediment model to incorporate up to four types of sediment. The extension was accomplished by modifying the sediment continuity and sediment transport equations formulated earlier for the single-sediment model. The extended formulas and their corresponding discrete numerical expressions take into account not only the capacity of each sediment type to be transported by a flow of given hydraulic characteristics, but also the interaction between types of sediment.

Graphic representation is also more complicated in the multisediment model, but the graphics represent the composition of simulated deposits as well as their age.

CHAPTER 8

Testing SEDSIM3

SEDSIM3 needs "battlefield" testing to determine whether it is flexible enough to simulate different sedimentary environments. The experiments that follow document SEDSIM3's versatility, and include applications to braided streams and marine deltas.

BRAIDED STREAMS

Braided streams differ from other streams in that they involve smaller proportions of water relative to the sediment transported. The lower proportion of water is a computational advantage because the number of fluid elements involved in flow calculations can be reduced. Scott (1986) simulated braided streams with SEDSIM3 to reproduce sedimentary features similar to those in the Ivishak Formation, which is a reservoir rock in the Prudhoe Bay oil field in the Arctic coast of Alaska.

Scott's experiments involve a rectangular area 3 km wide and 5 km long. Within the area there are three different sloping segments whose gradients decrease downstream. The uppermost segment is 1.5 km long and has a 4.6% slope, the middle segment is 2 km long and slopes at 3.0%, and the lower segment is 1.5 km long and slopes at 1.4% (Figs. 8.1 and 8.2).

Scott supplied "sources" of water and sediment at locations shown by crosses in Figure 8.2. During the experiments, the sources were activated in "pulses" at frequencies ranging from several times a day, to several times a year. During

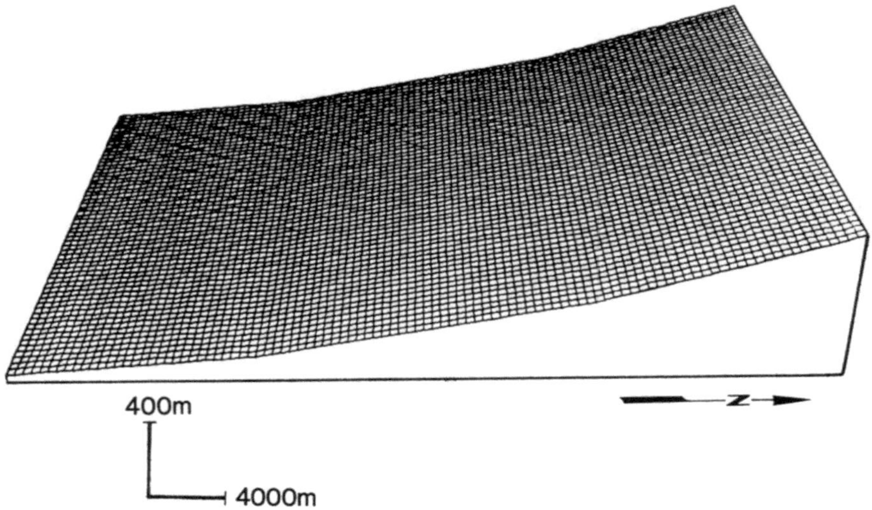

Figure 8.1 Initial topography for braided-stream experiments.

each pulse, a volume ranging from 80,000 to 120,000 m^3 of water and 100 to 1000 m^3 of sediment was released. Four types of sediment were supplied, with gravel constituting the smallest proportion, clay the highest, and sand and clay in intermediate proportions. Simulated time in individual experiments ranged from 10 to 500 years.

Figure 8.3 (page 129) illustrates Scott's first experiment, which involved 20 years simulated time, with graphic displays produced every 5 years. By the end of the first five-year interval, elongate bars had developed in response to channel shifting and braiding. Subsequently, the bars continued to change in shape and location, but their lengths, widths, and thicknesses did not change significantly. The persistent general form of the bars and channels suggests that equilibrium with respect to stream gradient, flow volume, and sediment volume was attained early in the experiment.

A second experiment by Scott involved the same area (Fig. 8.4*a*, on page 130), but differed in that fluid containing a higher proportion of gravel and sand was supplied via a single source located near the upper end of the area. As the experiment progressed, four channels separated by gravel bars were formed. The bars are shown in cross section (Time 1) in Figure 8.5 (page 131) and are labeled Area 1 through Area 4. Succeeding intervals (Times 2, 3, 4, and 5) are separated from each other by 5 simulated years, during which channels in Areas 1 and 3 filled with sand, while gravel filled most of the channel in Area 2. The channel occupying Area 4 received little sediment as late as Time 5.

Scott's third experiment involved a higher proportion of fine materials than in earlier experiments, with results shown in Figures 8.4*b* and 8.6. Figure 8.6 (page 132) represents sedimentary facies in two cross sections (GG′ and *HH*′) at five different times separated by intervals of 5 years. Section GG′ shows an overbank deposit that continued to migrate towards the left as the channel was progressively filled. Section *HH*′ reveals filling of an abandoned channel. The channel bed on the right side of *HH*′ consisted of sand until Time 3, whereas silt was deposited during an interval of reduced flow during Time 4. Flow through the channel increased during Time 5, depositing sand and burying the abandoned channel deposit formed earlier.

SCOTT'S SENSITIVITY TESTS

Scott performed "sensitivity" tests in which selected changes in input parameters were contrasted with other flow properties as the braided-stream deposits were formed. Measurements were made at locations denoted by bold dots in Figure 8.2. The initial test involved a "base case" with parameters listed in Table 8.1*a* (page 133). SEDSIM3 was allowed to run until flow and sediment transport parameters were more or less stable, and then the experiments were repeated several times. During each repetition, an input parameter was increased by 50% with respect to its value in the initial or base case, while other input parameters remained unchanged. Table 8.1*b* (page 133) shows responses to changes in flow velocity, depth, sediment discharge, and slope.

The sensitivity tests reveal SEDSIM3's responses to changes in initial slope in a braided-stream environment. For example, if slope is increased 50%, velocity increases, but depth also increases due to narrowing of the channels, and sediment discharge increases slightly. Downstream, however, increased deposition partly compensates for the initial increase in slope by causing the slope to decrease, thus returning to an equilibrium condition.

Response parameters are difficult to predict because of interactions between flow and sedimentation. For example, increasing the flow rate brings about immediate increases in velocity and in depth of flow in fixed channels. But, the increase in flow permits more sediment to be carried, resulting in greater erosion near the sources and causing the channels to become narrower, further increasing velocity and depth. Thus, the responses are complex and may not be foreseen. While the overall responses of braided-stream systems to changes in input conditions could be predicted with conventional formulas, the forms of migrating bars and channels would be difficult to forecast without a process

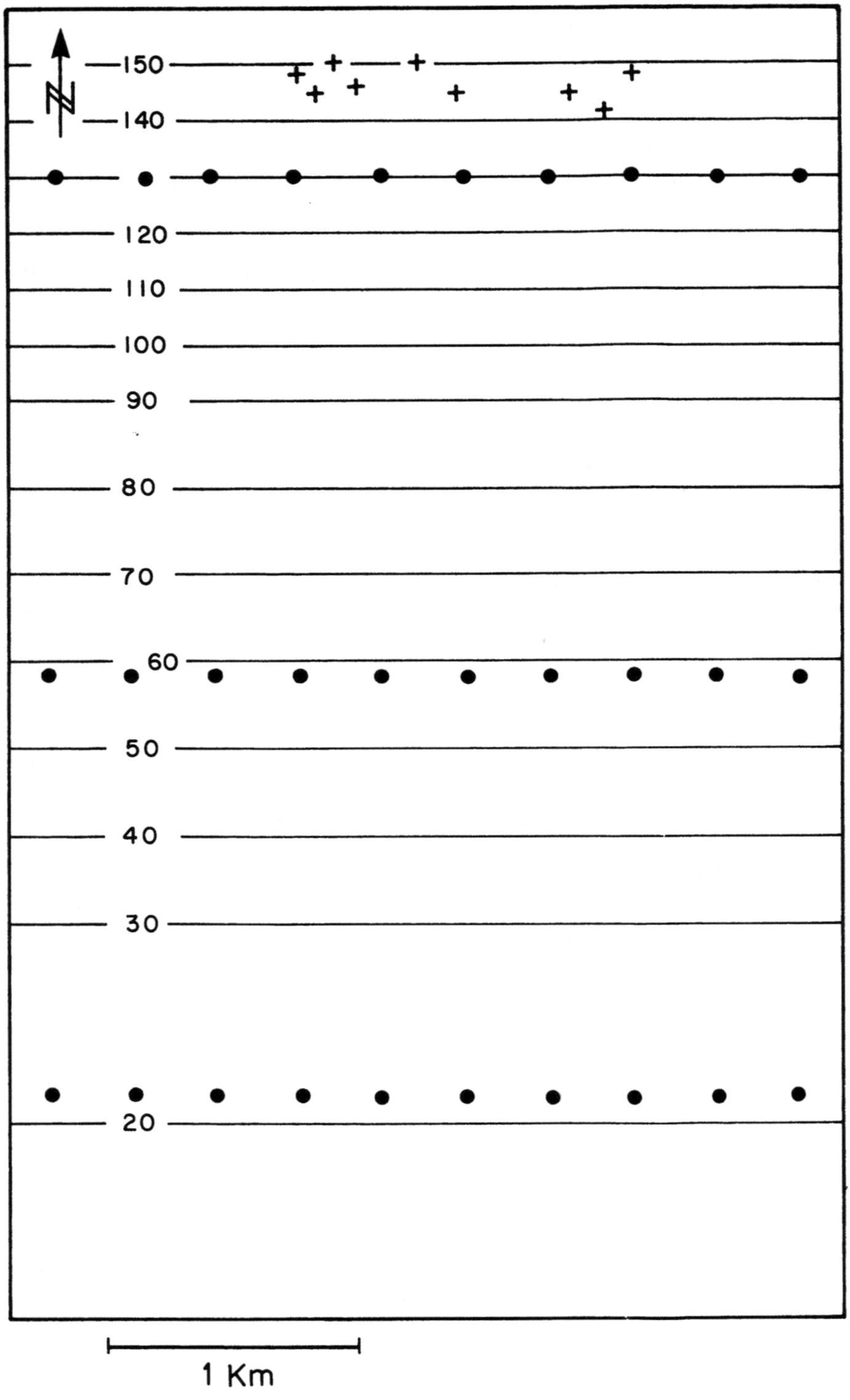
150
140
120
110
100
90
80
70
60
50
40
30
20
1 Km

simulation model such as SEDSIM3. Thus, simulation experiments are useful in predicting the responses to changes in flow rate, quantity of sediment, and slope.

Scott studied effects of variations in resistance to erosion on sediment transportability. Increasing the coefficient of erosion of basement has much the same effect as an increase in sediment input, because erosion of the basement provides more sediment for transport. Increases in the basement erodibility are accompanied by decreases in flow velocity, water depth, and sediment discharge, with little change in slope.

The complex interdependence between variables is also emphasized by SEDSIM3's responses when changes in sediment transportability occur. For example, when transportability increases, responses are similar to those accompanying a decrease in sediment input, with decrease in the flow's sediment content and an increase in erosion. Or, we may find that sediment discharge may decrease temporarily when sediment input is increased, because increased deposition upstream may be accompanied by an overall decrease in velocity.

Scott's experiments thus document SEDSIM3's use in analyzing fluvial environments. Physical laboratory experiments involving flumes and sand tables may yield predictions of responses of one or two variables to changes in input parameters when other parameters remain constant, but SEDSIM3 predicts the responses of a large complex fluvial systems, such as braided streams, by allowing entire systems to respond to changes in input conditions.

DELTAS

Deltas represent a complex of sedimentary environments that link land and sea and involve deposition in streams, lakes, beaches, flood plains, swamps, and in various marine environments. One of the challenges is to adapt SEDSIM3 to simulate such a complex of depositional environments, including the transitions between environments.

Several of our experiments involving deltas utilized an area 150 × 150 km that forms a simplified representation of part of the Gulf of Mexico in Miocene or Pliocene time. Figure 8.7 (page 134) shows initial topography in the experiment, which includes a narrow onshore area at the northern edge of the region and a sloping submerged platform extending 120 km offshore, reaching a maximum depth of 250 m within the area. Each experiment spanned 10,000 years simulated time.

In the first experiment, a single source of fluid and sediment representing the ancient Mississippi River was located onshore, providing a steady inflow of

Figure 8.2 Contour map of initial topography, corresponding to Figure 8.1. Slope progressively steepens toward north. Contours in meters. Sources of water and sediment are indicated by crosses, while bold dots indicate locations where flow, slope, and sediment content were measured in sensitivity analyses.

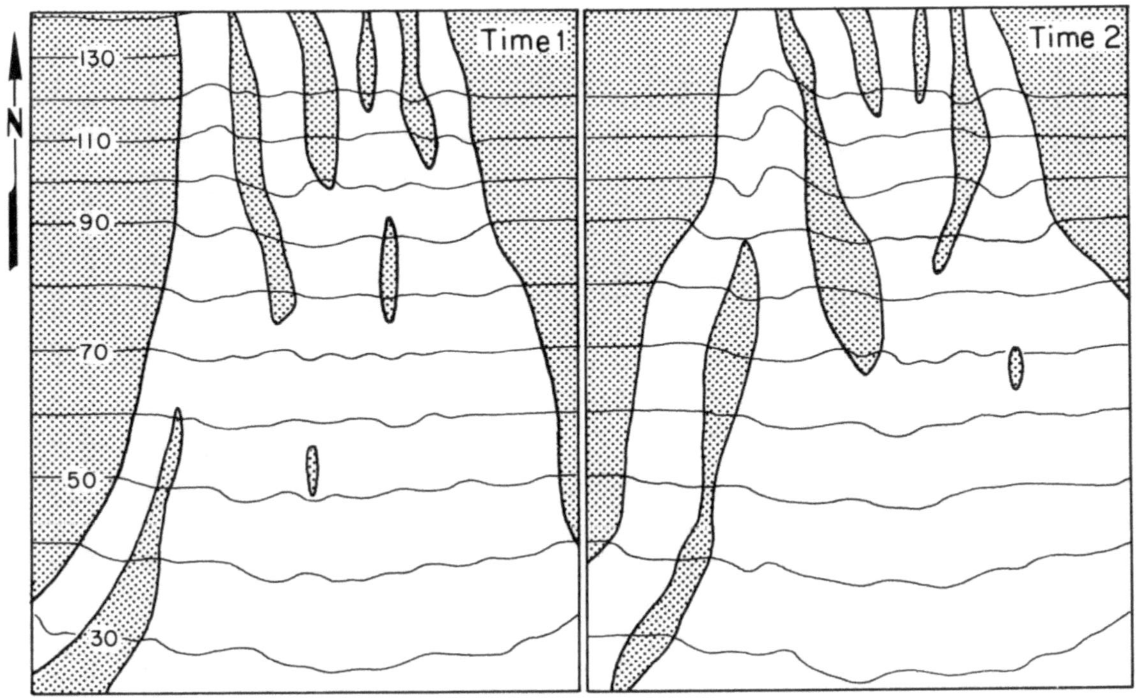
Time 1
Time 2
N
130
110
90
70
50
30

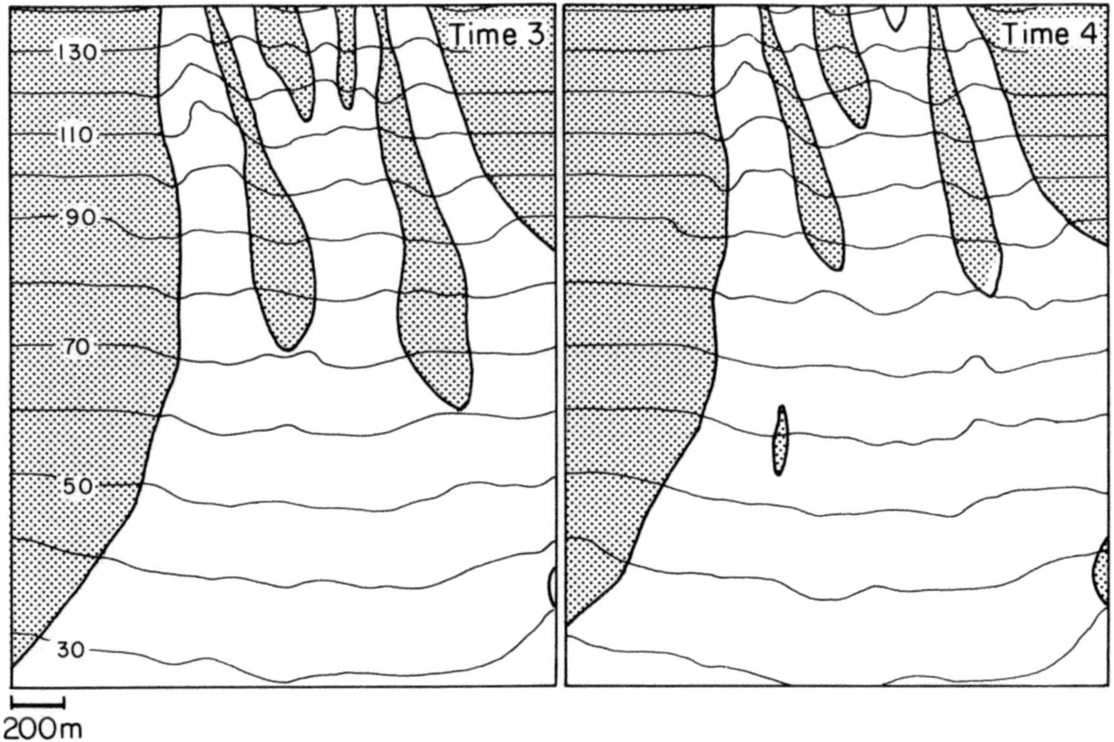

Figure 8.3 Maps showing successive changes in braided streams at intervals of five years (Times 1, 2, 3, and 4). Bars are shown with stippled pattern. Channels occupy areas between bars. Contours in meters.

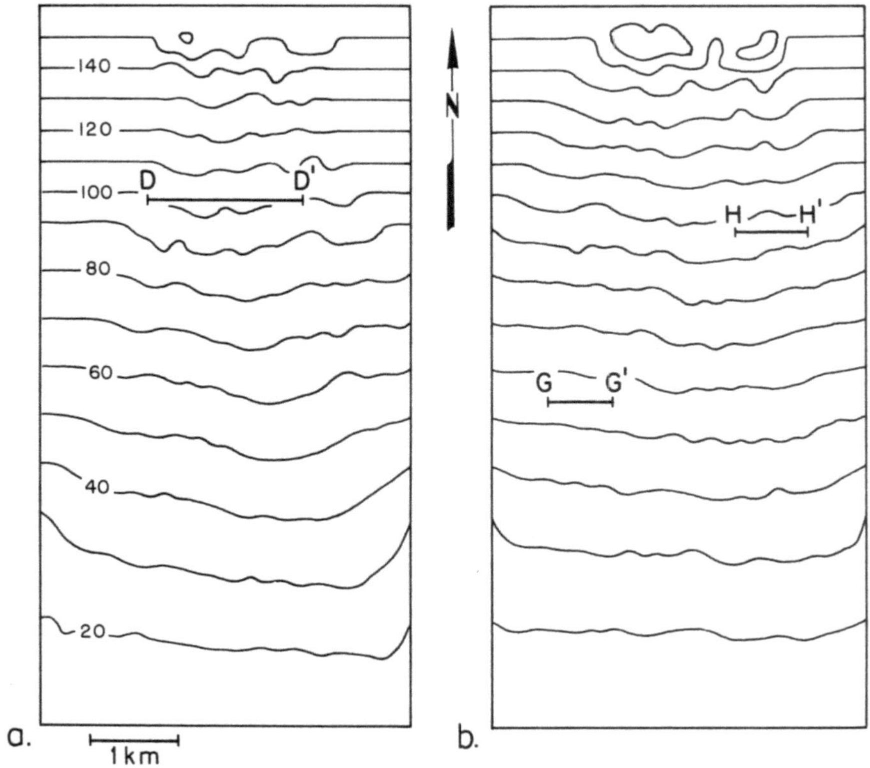

Figure 8.4 Topography produced in two braided-stream simulation experiments; contours are in meters. (*a*) Experiment involving development of bars and channels (section D-D′ is shown in Figure 8.5). (*b*) Experiment involving development of overbank deposits and abandoned-channel deposits (sections G-G′ and H-H′ are shown in Figure 8.6.)

7500 m^3/s of water. The river had a sediment concentration at the source of 0.002, providing about 30,000 kg/s of sediment that consisted of 15% medium sand, 22% fine sand, 28% silt, and 35% clay. As time passed, there were significant changes. The topographic configuration after 10,000 years is shown in Figure 8.8 (page 135). During this interval, the shoreline prograded about 25 km, with several channels forming onshore, and a complex of distributary lobes formed offshore.

Effective displays of the results are provided by two kinds of vertical sections. Sections that represent ages are labeled with different colors for every 100-year interval (using the sequence of colors shown in Plate 1*a*). Sections that represent sediment composition (facies sections) use colors to show pro-

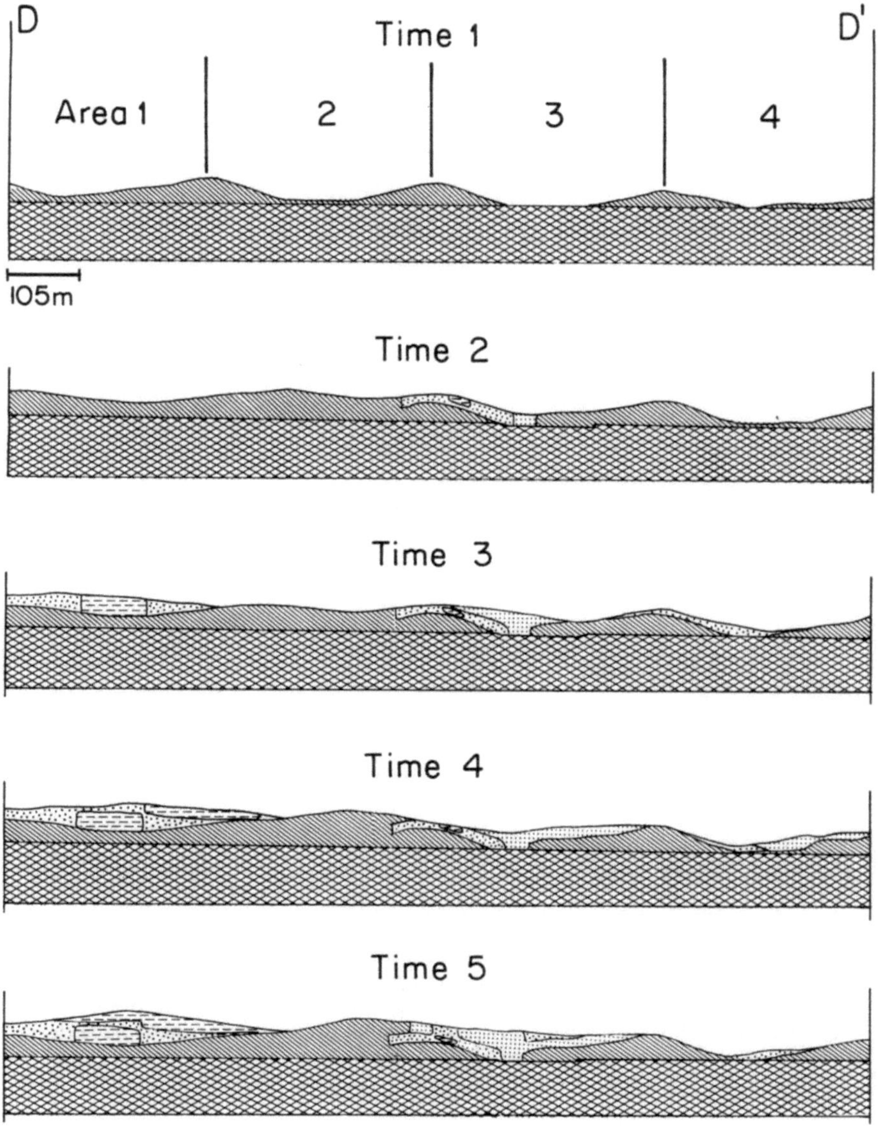

Figure 8.5 Sections at intervals of five years showing braided-stream deposits. Vertical exaggeration = 10. Trace of section is shown in Figure 8.4*a*. Coarser deposits are stippled.

portions of medium sand, fine sand, silt, and clay, which are represented in pure form by red, green, blue, and black, respectively, with mixtures represented by color combinations shown in the scheme in Plate 1*b*.

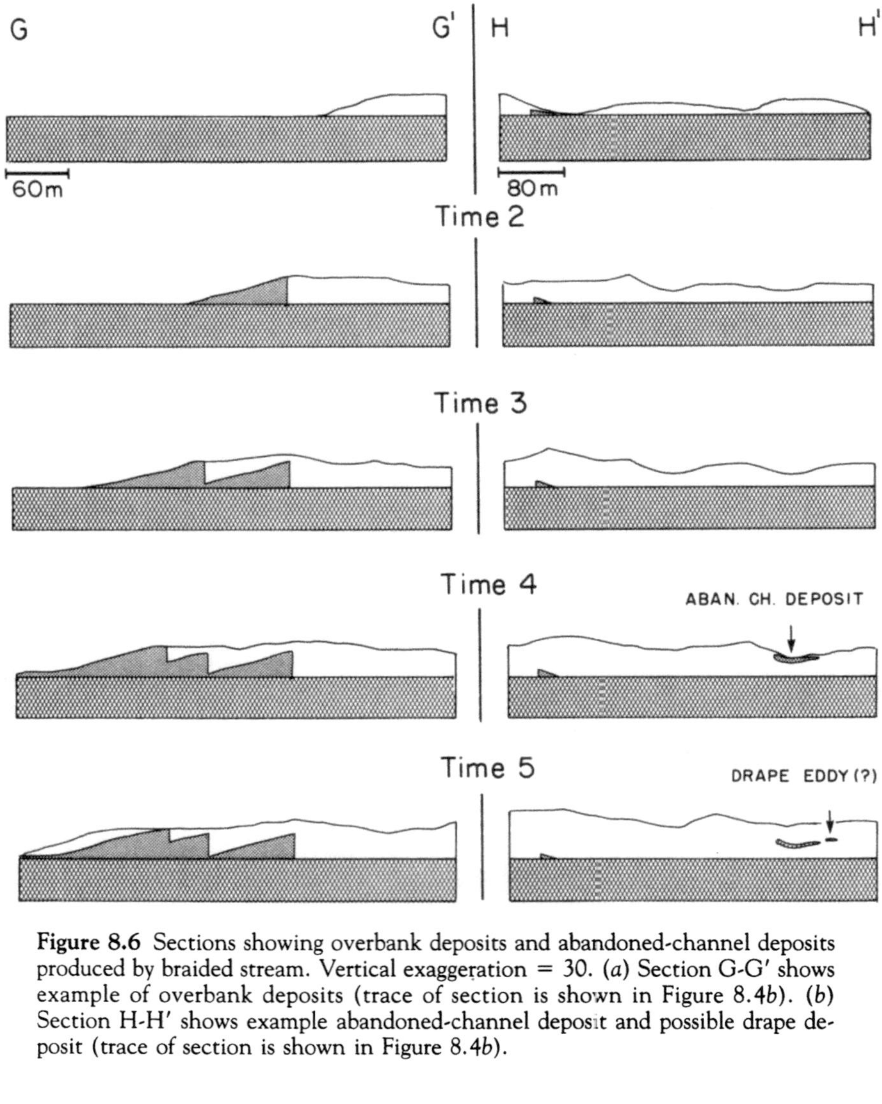

Figure 8.6 Sections showing overbank deposits and abandoned-channel deposits produced by braided stream. Vertical exaggeration = 30. (*a*) Section G-G′ shows example of overbank deposits (trace of section is shown in Figure 8.4*b*). (*b*) Section H-H′ shows example abandoned-channel deposit and possible drape deposit (trace of section is shown in Figure 8.4*b*).

The sections show major variations from onshore to offshore. Plates 2, 3, and 5 are north-south sections perpendicular to the shore, with sections shown every 2000 years. Toward shore, channels were repeatedly cut and filled over the 10,000-year span, whereas deposition predominated offshore. Facies sections in Plates 2, 3, and 5 document progressive decreases in grain size with distance offshore. The beds thin moving away from shore, with large variations

TABLE 8.1 Results of Sensitivity Analyses in Braided-Stream Experiments

	(*a*) *Fluvial Parameters for Base Case*			
	Velocity (*m/s*)	*Depth* (*m*)	*Sediment Discharge* (m^3/s)	*Slope*
	0.140	0.107	0.225	0.035
	(*b*) *Resulting Change with Respect to Base Case*			
Variable increased 50%	*Velocity* (%)	*Depth* (%)	*Sediment Discharge* (%)	*Slope* (%)
Flow rate	15	80	323	−2
Initial slope	15	9	60	33
Basement erosion factor	−9	−11	−15	−6
Sediment transportability	11	257	97	4
Sediment input	46	49	53	13

Source: Data provided by Scott (1986).
Note: Part *a* lists base case values for velocity, depth, sediment discharge rate, and slope. Part *b* expresses changes in each of these four values, expressed in percent, when each of other variables (flow rate, initial slope, basement erosion factor, sediment transportability, and sediment input) were individually increased by 50%.

in composition even though the source of fluid and sediment remained constant. The variations result mostly from lateral migration of distributary flows moving along the bottom that were shunted from side to side as they sought locations to deposit sediment, either temporarily or permanently.

The gross forms of the beds are "oblique progradational," in the terminology of Sangree and Widmier (1977), and Mitchum (1977), and consist of beds that terminate by toplap or truncation updip. The complex of beds is typical of deltaic deposits and contains numerous channel deposits toward shore, reflecting an environment of strong fluvial currents, rapid influx of sediment, and constant sea level. The toplapping relationships involve earlier nearshore deposits that were later eroded and truncated by fluvial currents.

Sections parallel to shore also show lateral variations. East-west sections in Plates 4 and 6 show lateral shifting of coarse and fine materials, much as might be interpreted from seismic sections through deltaic deposits. If compaction, isostatic compensation, and sea level rise had been incorporated in the experiment, there probably would have been much less truncation of near-shore deposits, which would tend to be preserved by burial.

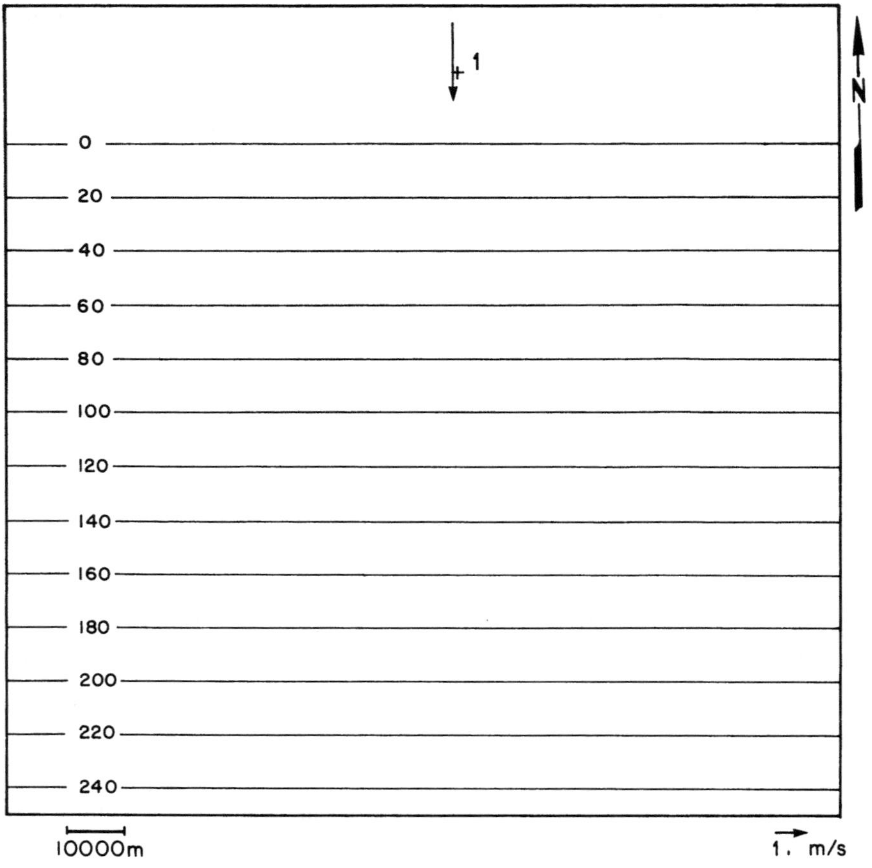

Figure 8.7 Map of submerged topography at outset of experiment involving simulation of large delta complex. Medium sand, fine sand, silt, and clay were supplied by river that flowed continuously and entered area at location denoted by cross. Contours are in meters below sea level.

RESPONSES TO CHANGES IN SEDIMENT COMPOSITION

Changes in slope, flow rate, sediment content, and sediment composition can cause major changes in simulated deltaic deposits. We have not undertaken a complete sensitivity analysis, but the responses to changes in sediment composition and flow rate were studied in three subsequent experiments. The first of the subsequent experiments assumes a smaller proportion of sand and a lower flow rate (Plate 7), whereas the second assumes a higher proportion of

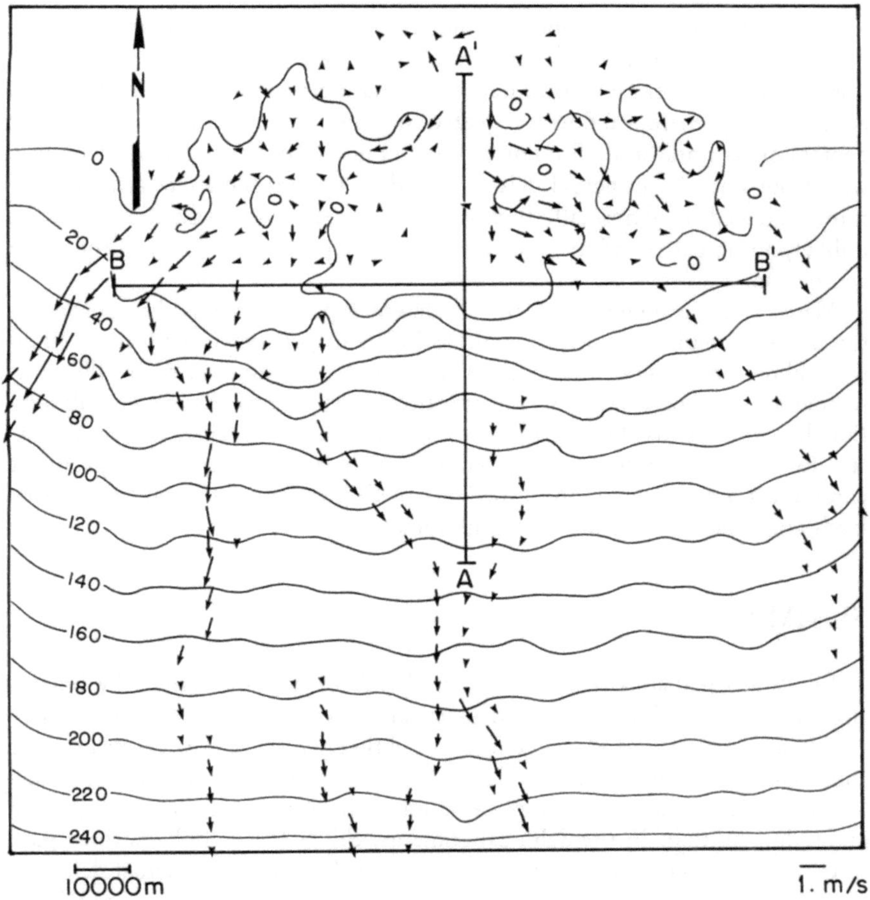

Figure 8.8 Topography of deltaic deposits and flow velocities after 10,000 simulated years. Sections A-A′ and B-B′ are shown in Plates 2, 3, 4, 5, and 6.

sand and a higher flow rate (Plate 8). If we compare Plate 3, representing the initial experiment, with the two subsequent experiments, we find major differences in the geographic distribution of grain sizes. Sand has been deposited closer to shore in the experiment shown in Plate 7 and farther from shore in the experiment in Plate 8. Also, the slopes of prodelta deposits progressively decrease in the sequence represented by Plates 7 and 8. The third subsequent experiment (Plate 9) has the same flow rate as that shown in Plate 7, but it has the same proportion of sand as the experiment shown in Plate 8.

The experiment in Plate 7 involves finer sediment and reveals a pronounced "source effect" where the river entered, where a deep depression was eroded, with extensive channels forming near shore. The erosion and channeling are related to the lower proportion of coarse sediment, which increased the flow's capacity to erode earlier deposits. Channels commonly form in actual environments where the flow is depleted in coarse sediment, although the deep depression is unnatural. Similarly, the "mound" near the source in Plate 8 would not form in an actual delta. Mounds and depressions are artificial features produced by abnormally high or low sediment loads, which are virtually unlikely in actual rivers that quickly adjust their sediment loads as they seek equilibrium.

Plates 7 and 8 demonstrate that sediment input loads must be appropriate for given hydraulic conditions if SEDSIM3 is to perform realistically. Otherwise, abnormal erosion or deposition may occur near sources where fluid and sediment enter and flows have not traveled far enough to achieve equilibrium. But the experiments document that SEDSIM3 behaves realistically if parameters for flow are appropriately selected. Furthermore, even if inappropriate parameters are provided, the flows in SEDSIM3 eventually stabilize so that loads and fluid volumes are in equilibrium.

SUMMARY

SEDSIM3 realistically simulates the details of deposits formed by braided streams and on deltas. The braided-stream experiments illustrate how SEDSIM3 can be used to study the effects of subtle changes in flow conditions and the stream's sediment content. SEDSIM3 has an advantage over simple formulas in that it considers the complicated interactions between topography, flow, and sedimentation. In the delta experiments, SEDSIM3 reproduces the typical forms and the compositional variations frequently observed in actual deltas, as viewed both in longitudinal and transverse sections. Furthermore, SEDSIM3 also responds to changes in flow conditions and in sediment supply, in the same manner as actual deltas.

CHAPTER 9

SEDSIM3's Application to Two Real-World Problems

SEDSIM3 can operate over a range of scales and can be focused on various applications, including prediction of sedimentary facies that may serve as oil and gas reservoirs. Two real-world examples follow. The first involves Simpson Canyon, an ancient submarine canyon buried beneath Alaska's Arctic coast. The second involves Tertiary deltaic deposits in the Golden Meadow oil field of coastal Louisiana.

SIMPSON CANYON

Simpson Canyon lies beneath the National Petroleum Reserve in Alaska (NPRA, Fig. 9.1) and lies east of the giant Prudhoe Bay oil field. It has been mapped from continuous seismic sections that intersect it at different angles. Its history involves repeated stages of carving and filling (Tetzlaff and Harbaugh, 1985). It is possible that Simpson Canyon contains oil and gas, but it has been penetrated by only one or two exploratory wells, and thus remains essentially unexplored by drilling, particularly considering that its volume may be as great as 200 mi^3 (it is about 25 miles long, 10 miles wide, and 4000 feet deep). The main question is whether Simpson Canyon contains reservoir rocks that could yield oil and gas.

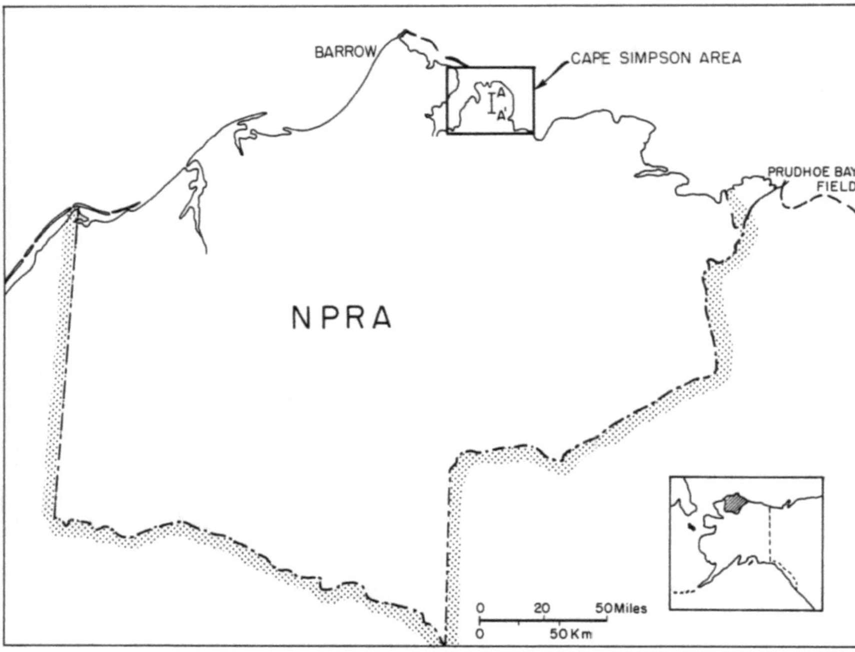

Figure 9.1 Index map showing location of Cape Simpson in northern part of National Petroleum Reserve in Alaska (NPRA).

Geologic Setting

The canyon is both carved in and filled with Cretaceous siltstones and graywackes that form part of the Brookian sequence (Carter et al., 1977). Deposition of the Brookian sequence involved uplift of an ancestral Brooks Range to the south, accompanied by rifting of the Artic Ocean, providing both a source of sediment and a place for the sediment to be deposited in the form of a thick deltaic sequence. Simpson Canyon's history is closely linked with these tectonic events.

The pebble shale is the earliest deposit in the Brookian sequence and consists of 300 to 600 feet dark gray sandy shale. The pebble shale may be an important source rock for hydrocarbons in the region, including those in the Prudhoe Bay, Umiat, and Point Barrow fields. Unconformably over the pebble

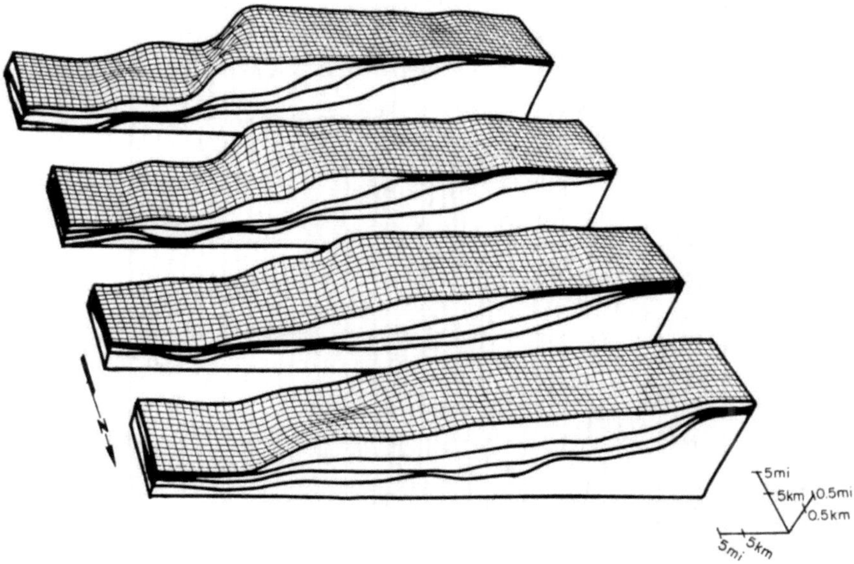

Figure 9.2 Paleotopographic reconstruction of deltaic deposits before Simpson Canyon was formed. Four paleotopographic surfaces are shown and are based on information provided by intersecting seismic sections.

shale there are thick wedges of sediment derived from a southerly source and that form the Torok Formation and the Nanushuk Group. The Torok and the Nanushuk partly intergrade laterally and represent a large delta complex whose deposits are sigmoidal in cross section (Fig. 9.2) and prograde toward the northeast, as shown in seismic sections. The Torok represents bottomset and foreset beds of the delta complex, and the Nanushuk represents the topset beds. The Nanushuk is covered by beds of the Colville Group of late Cretaceous age, which consists mostly of marine shales. Part of the material that fills Simpson Canyon consists of beds in the Colville Group.

We can distinguish three ancient erosion surfaces in Simpson Canyon, each of which marks a stage of canyon carving and subsequent refilling. The lowest and oldest surface (Figs. 9.3, 9.4) has several smaller tributary canyons that may have been carved by subaqueous erosion, or alternatively may represent large-scale slump deposits. The floor of the main part of the canyon is relatively flat and demarks the lowest level to which erosion lowered the canyon's

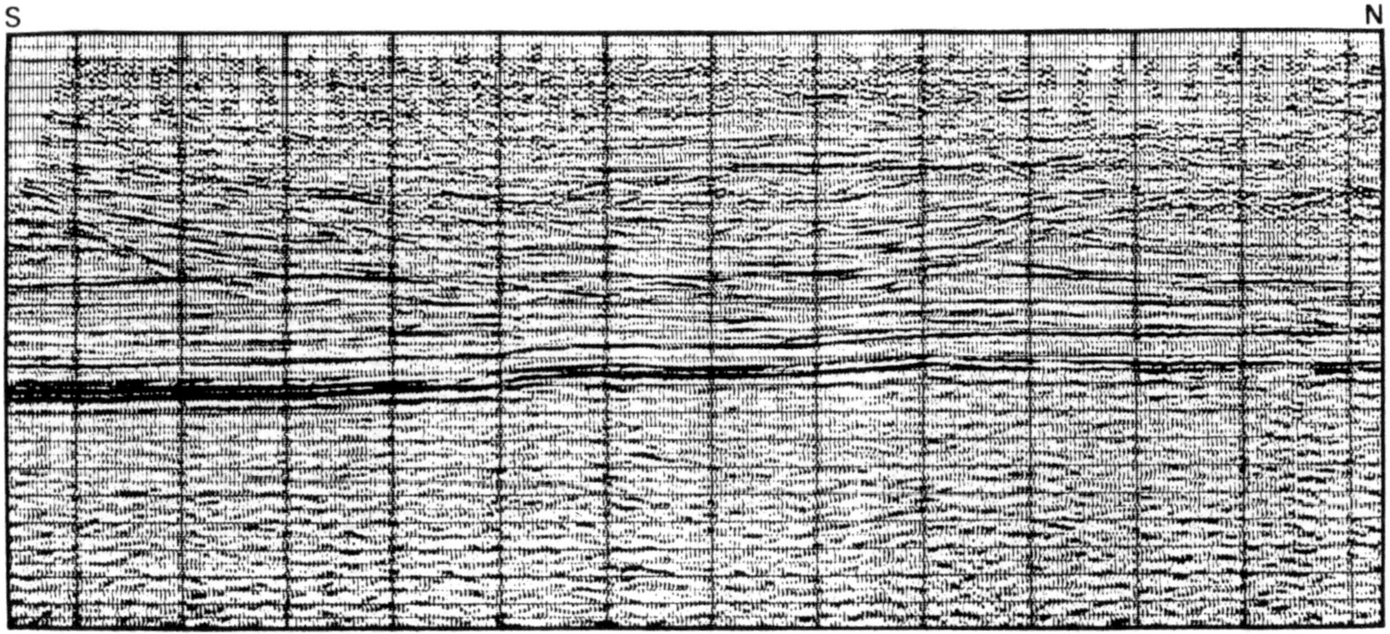

a

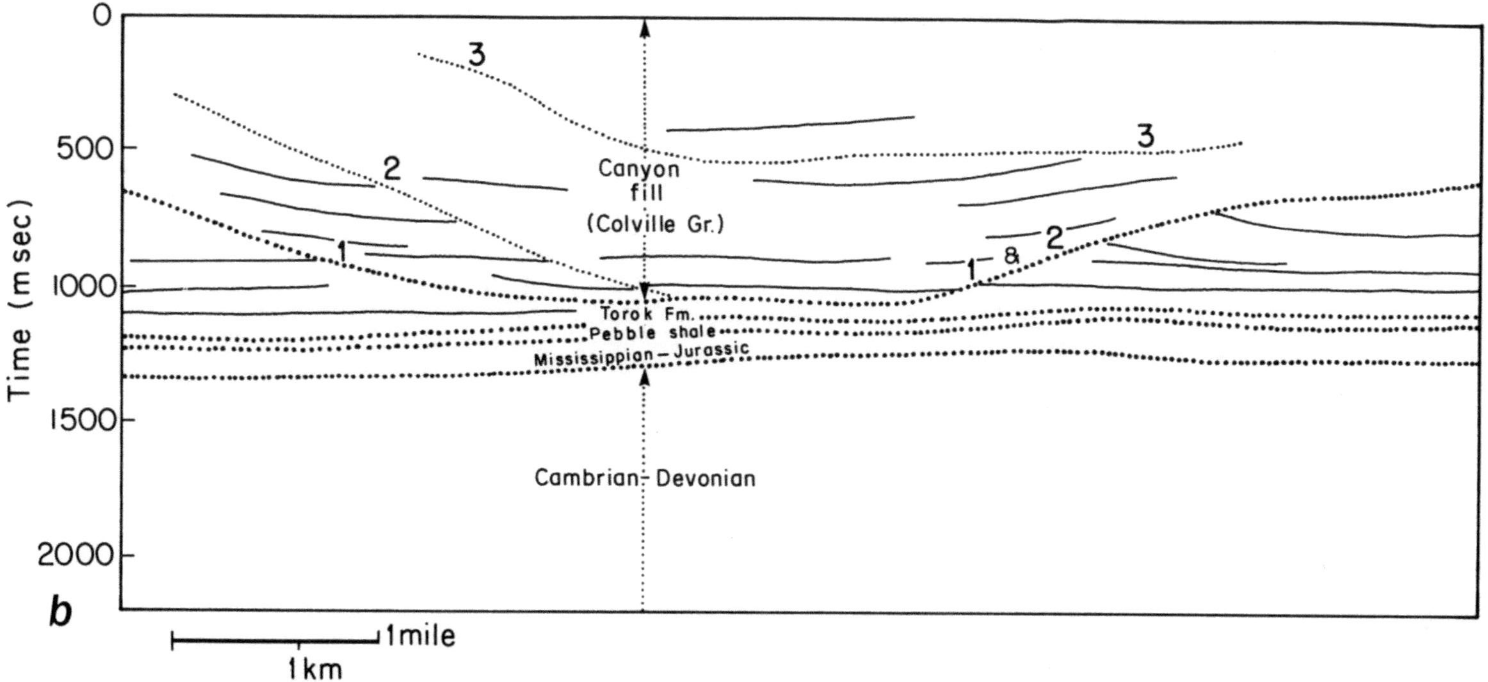

Figure 9.3 Cross section A-A′ through Simpson Canyon, whose trace is shown in Figure 9.1. Section represents horizontal distance versus seismic two-way travel time in milliseconds. (*a*) Seismic section whose trace is shown in Figure 9.1. (*b*) Outline of features interpreted in seismic section in which numbered lines denote three former submarine erosion surfaces.

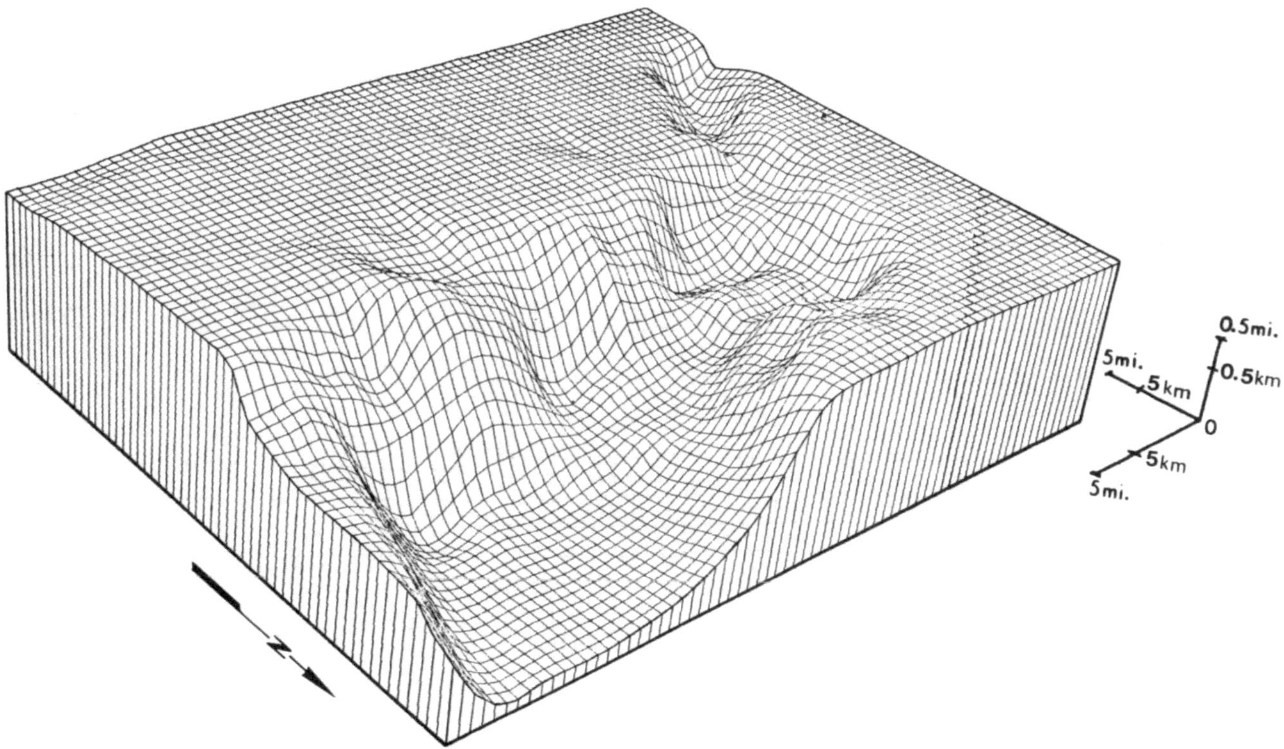

Figure 9.4 Lowermost submarine erosion surface (horizon 1 of Figure 9.3*b*) of Simpson Canyon.

floor, coinciding with the base of the Torok Formation. Perhaps further downcutting was inhibited by relatively resistant qualities of the pebble shale, but it seems likely that the canyon floor's former elevation was controlled by local erosional base level governed by the elevation of the seafloor beyond the canyon's mouth.

Carving the Canyon

Since the floor of Simpson Canyon lies more than 4000 feet below the continental shelf into which it is carved, it is unlikely that Simpson Canyon could have been carved by subaerial erosion. Sea level would have had to be reduced about 4000 feet if the canyon were to have been carved by subaerial erosion. Uplift of the Barrow arch might have had some effect, but there is no evidence of widespread subaerial erosion that could have accompanied such an uplift. Thus, we turn to turbidity currents as the agent of erosion. Turbidity currents have long been regarded as agents capable of carving and filling submarine canyons, and they are important in forming widespread deposits on continental shelves and slopes and in the deep sea. We have assumed that turbidity currents were largely responsible for carving and filling Simpson Canyon, and we adapted SEDSIM3 to represent turbidity flows capable of eroding and filling features of the magnitude of Simpson Canyon.

We interpreted the successive forms of Simpson Canyon by mapping elevations of the former erosion surfaces as estimated from seismic sections, and representing the surfaces with fishnet displays (Fig. 9.4) that provide topographic reconstructions at successive stages.

In our simulation experiments, we specified the initial topography as well as the type of sediment, the frequency and volume of turbidity flows, and the sediment concentration of the flows. The initial topography is shown in Figure 9.5, consisting of a gentle slope intersected by a steep scarp. The frequency of flows is problematical. Frequent turbidity flows over a short period might yield the same effect as less frequent flows over a longer period. But, in spite of uncertainty as to the frequency and magnitude of turbidity flows, the canyon's changing topographic form and changes in its volume of fill can be interpreted with greater confidence.

In our experiments we found that SEDSIM3 was not sensitive to the initial sediment concentrations, which quickly adjusted to prevailing conditions of flow, but we found that adjustments in flow volume are important. The best match between the canyon observed in seismic sections and our simulations is shown in Figure 9.6 (pages 146–147), which spans 60,000 years of simulated time. We assumed that turbidity flows recurred every 20 years on average, and that each flow had an initial volume of 500 million cubic meters of water con-

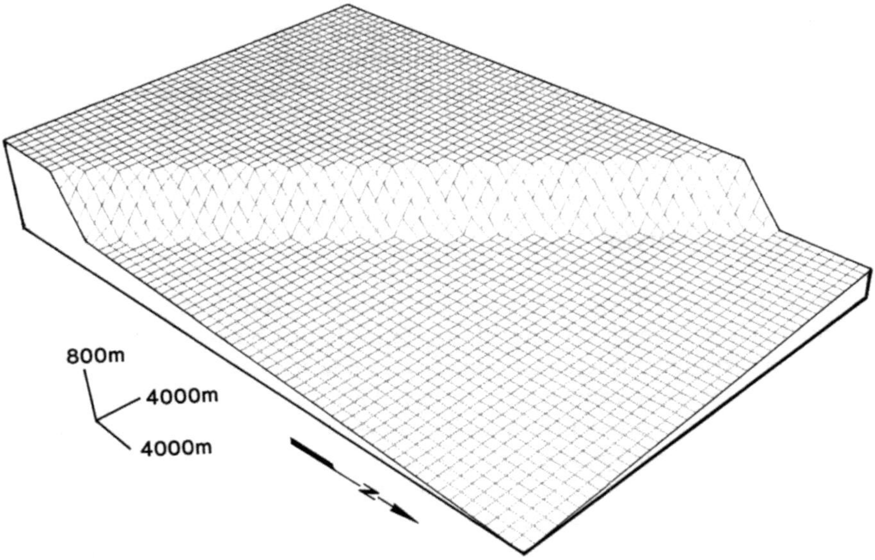

Figure 9.5 Initial submarine topography assumed in Simpson Canyon simulation experiment.

taining suspended silt, with the silt forming 20% of the flow by weight. The best accord was obtained after about 1000 such flows had occurred (Fig. 9.6*a*).

If our assumptions about the erodibility and transportability of silt in the Torok and the Nanushuk are reasonable, we then have a rough measure of the aggregate volume of turbidity flows needed to carve the canyon. Our experiments also yield predictions about the shape and areal extent of a depositional fan that may have formed beyond the mouth of the canyon (and unfortunately beyond the range of seismic sections available to us). Figures 9.6 and 9.7 show a relatively thin fan that slopes at a low angle and extends over a large area. If such a fan exists, it may be a potential oil and gas reservoir, although it may be too thin to be commercially important. Exploration of Simpson Canyon by additional drilling seems warranted, perhaps permitting our "predictions" to be tested.

GOLDEN MEADOW FIELD

The second real-world application involves deposits at Golden Meadow oil and gas field. Golden Meadow is situated in southern Louisiana (Fig. 9.8), in Lafourche Parish and within Louisiana's onshore Gulf Coast oil-producing region. Deposits in the region consist of a thick Cenozoic deltaic sequence that

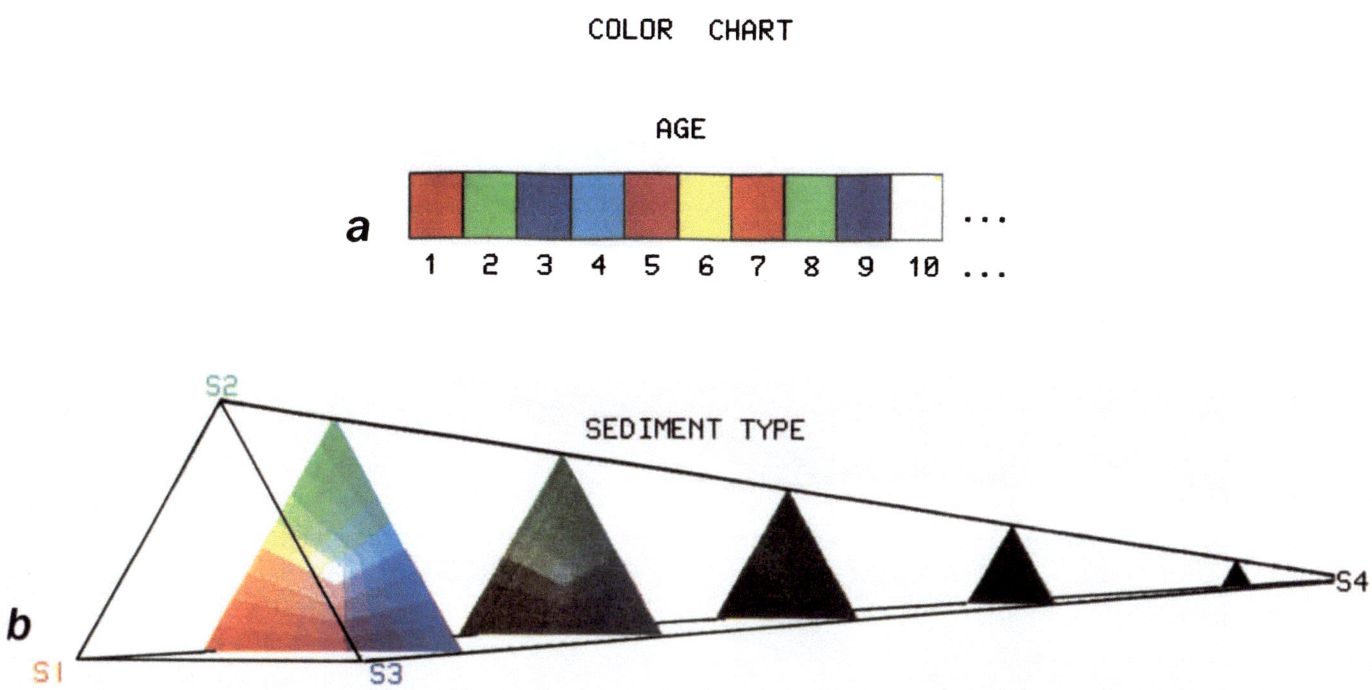

Plate 1 Index to colors used in plates. (*a*) Sequence of ten colors that define ages of sedimentary deposits shown in vertical sections. Sequence can be successively repeated to represent any number of deposits. (*b*) Tetrahedral diagram used to represent composition of sediments in vertical sections and facies maps. Colors at vertices of tetrahedron (red, green, blue, and black) represent pure sediment types, while additive color mixtures represent corresponding mixtures of sediment. White is substituted for black when displays are shown at color terminal and black background is used.

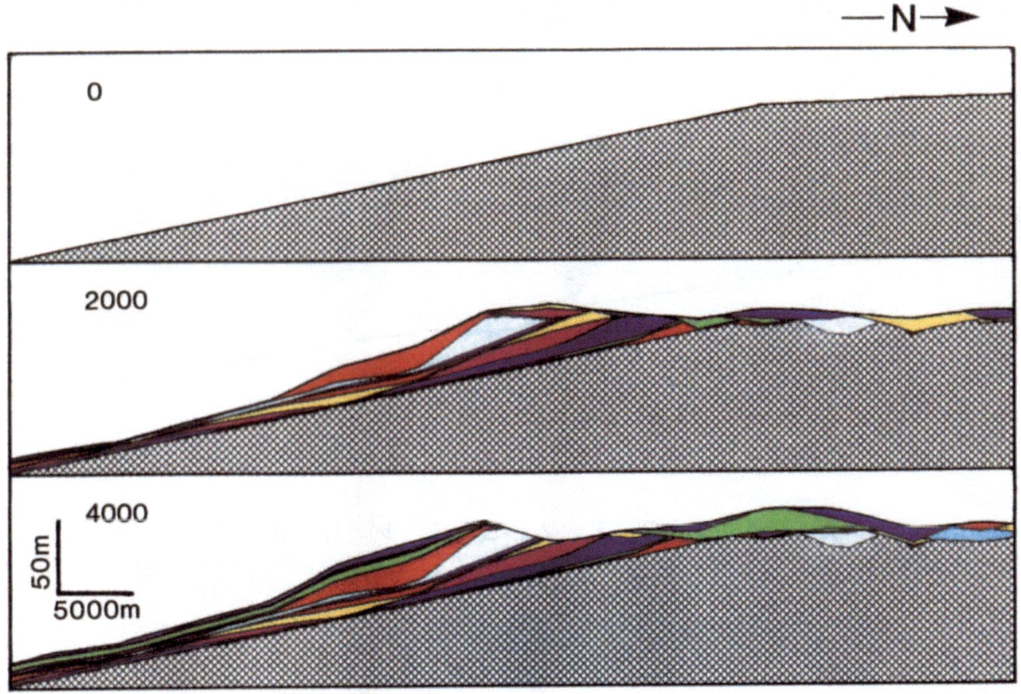
N
0
2000
4000
50m
5000m

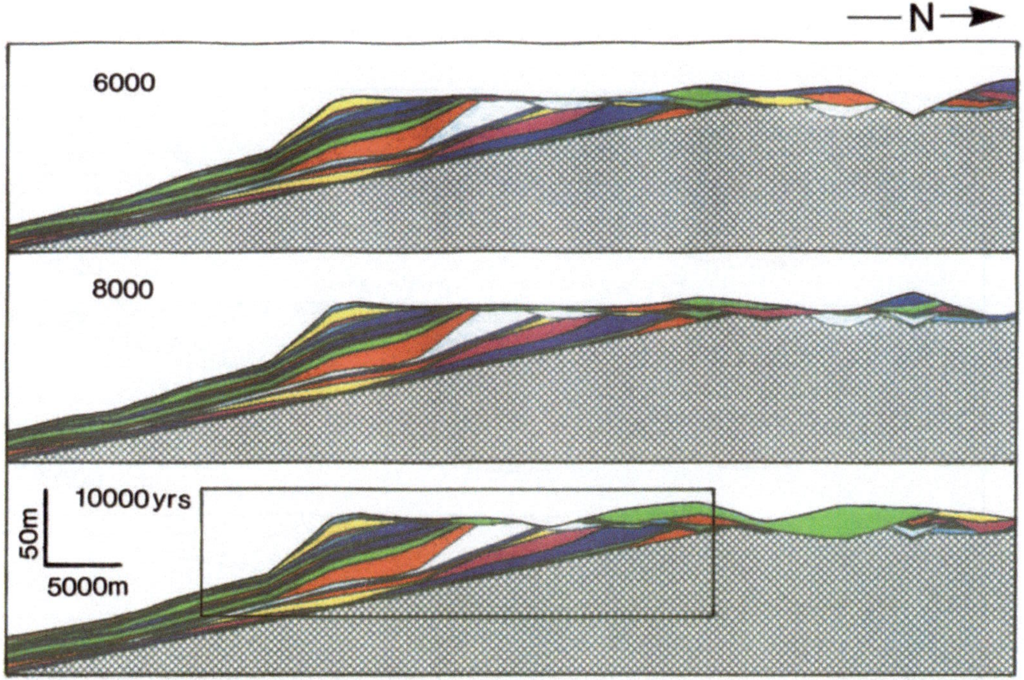

Plate 2 "Dynamic" display consisting of north-south section A–A′ that shows ages of deposits formed successively over span of 10,000 years. Each panel represents 2000 years. Box outlines enlargement shown in Plate 5*a*, and trace of section is shown in Figure 8.8. When color terminal is used, panels can be displayed at much shorter intervals and in rapid succession, providing dynamic video display.

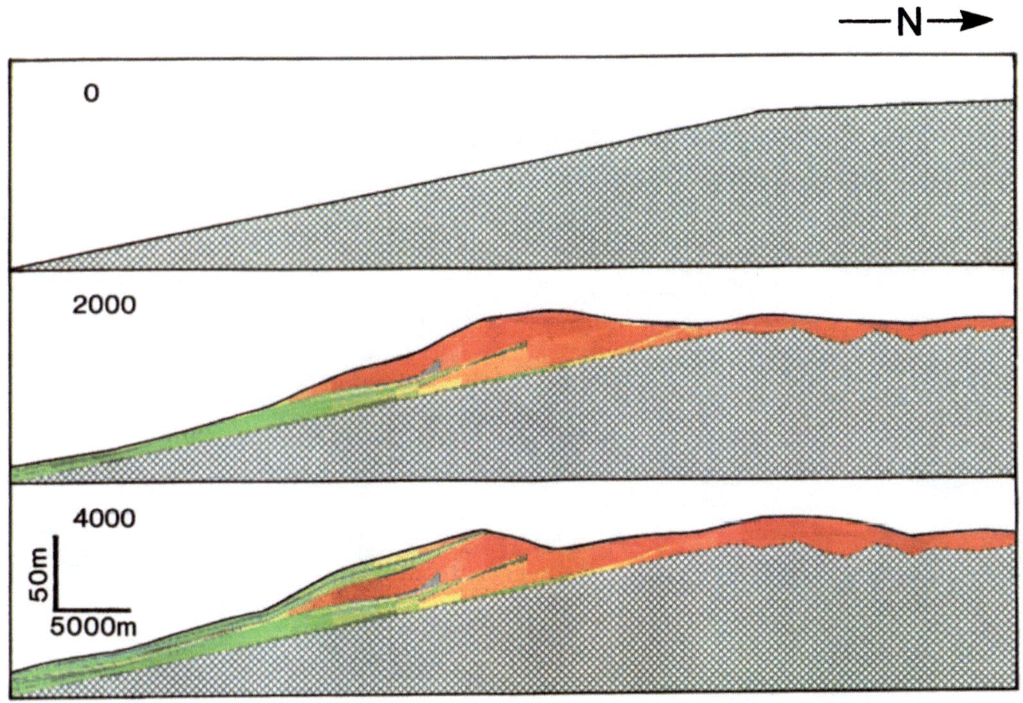
N
0
2000
4000
50m
5000m

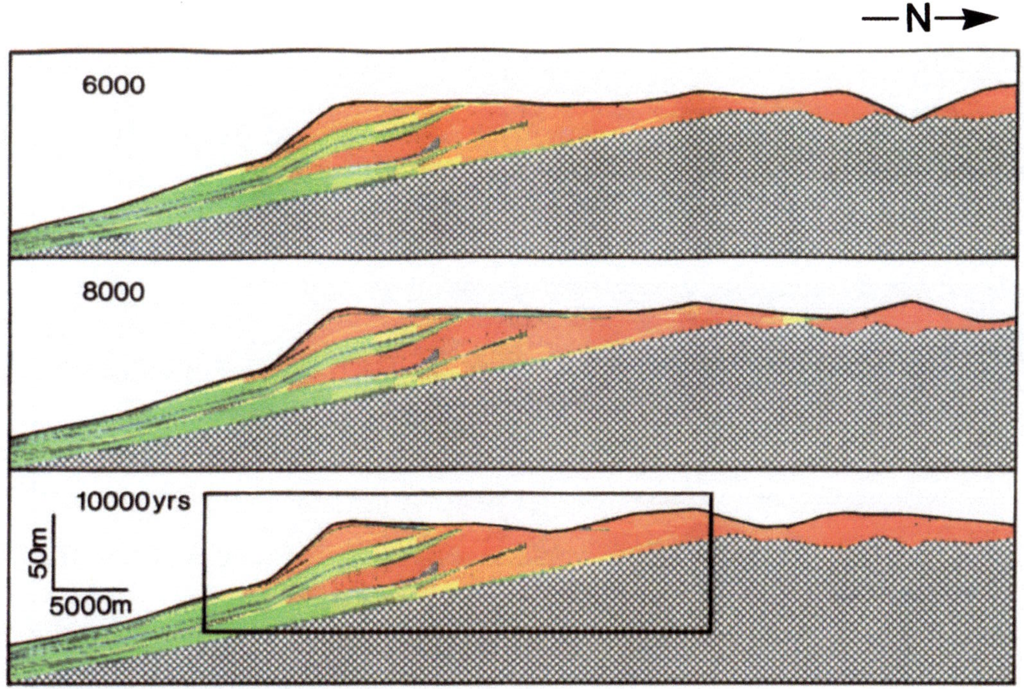

Plate 3 North-south section A–A′ showing composition of deposits at intervals of 2000 years, corresponding to age sections shown in Plate 2. Box outlines enlargement shown in Plate 5. Trace of section is shown in Figure 8.8.

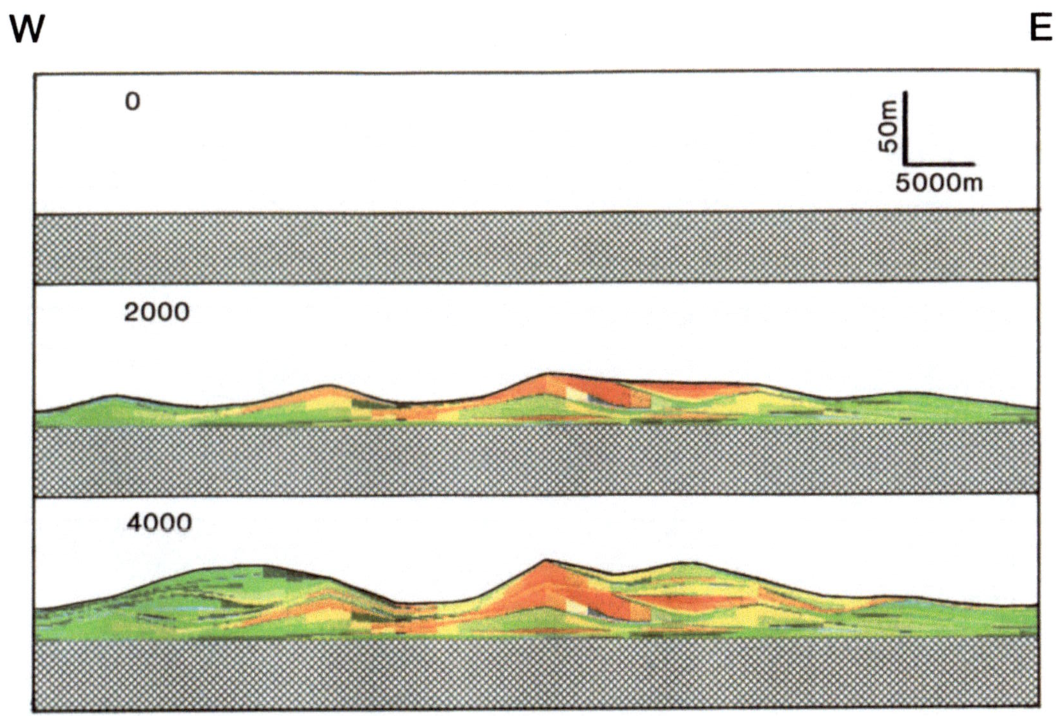
W
E
0
50m
5000m
2000
4000

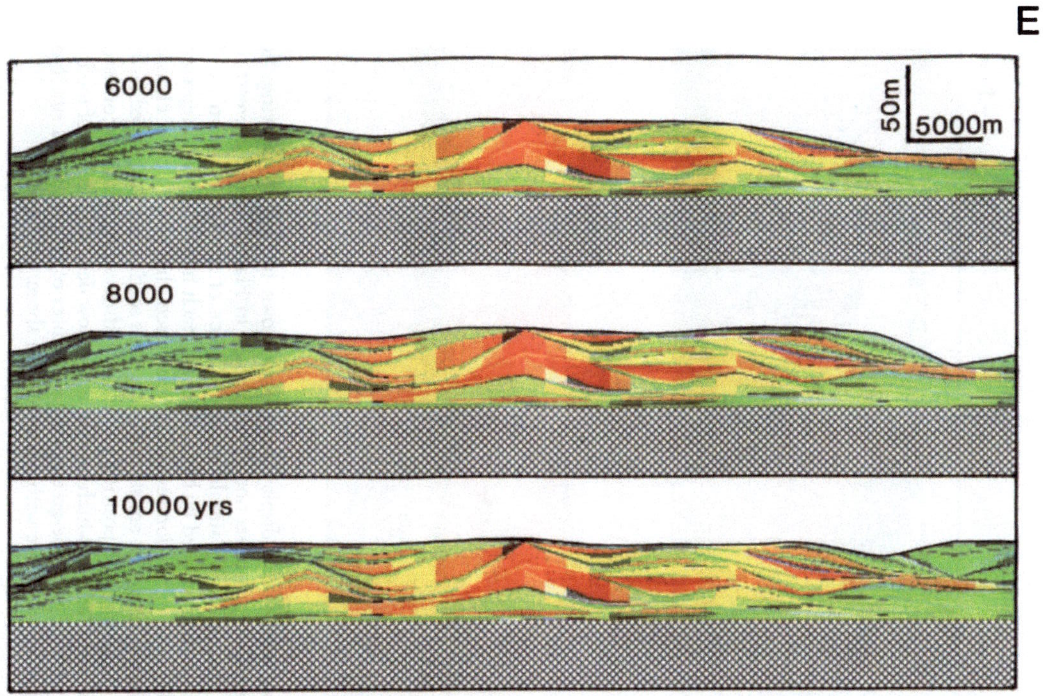

Plate 4 "Dynamic" east-west section B–B′ showing composition of deposits at intervals of 2000 years over span of 10,000 simulated years. Trace of section is shown in Figure 8.8. Colors reveal lenticular nature of deposits. Red and orange denote relatively coarse deposits formed by deltaic distributary current that shifted from side to side.

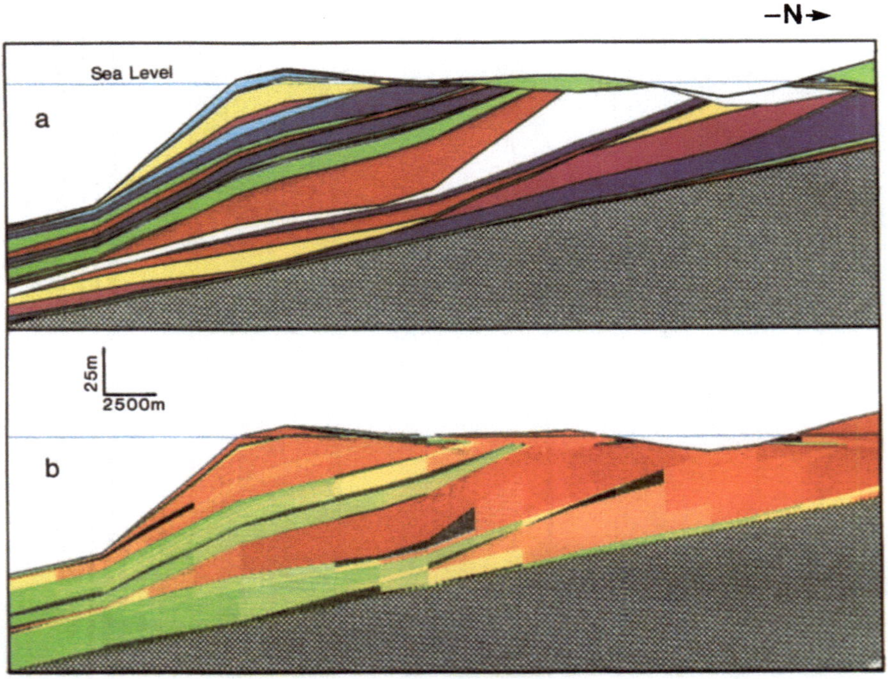

Plate 5 Example use of color display procedures shown in Plate 1: Drawings present an enlargement of part of section A–A′ shown in Plates 2 and 3 representing deposits formed during 10,000 simulated years. Trace of section is shown in Figure 8.8. (*a*) Ages of beds denoted by bands of color, each individual band represents 400 years of geologic time. Section reveals extensive scouring and filling of channels toward shore, whereas beds that are sigmoidal in cross section having formed farther from shore. (*b*) Sediment composition in which red denotes mixtures consisting mostly of medium sand, and in which orange and yellow denote mixtures of medium and fine sand, and blue and blue-green denote mixtures of fine sand and silt. Note progressive decrease in grain size with distance from shore.

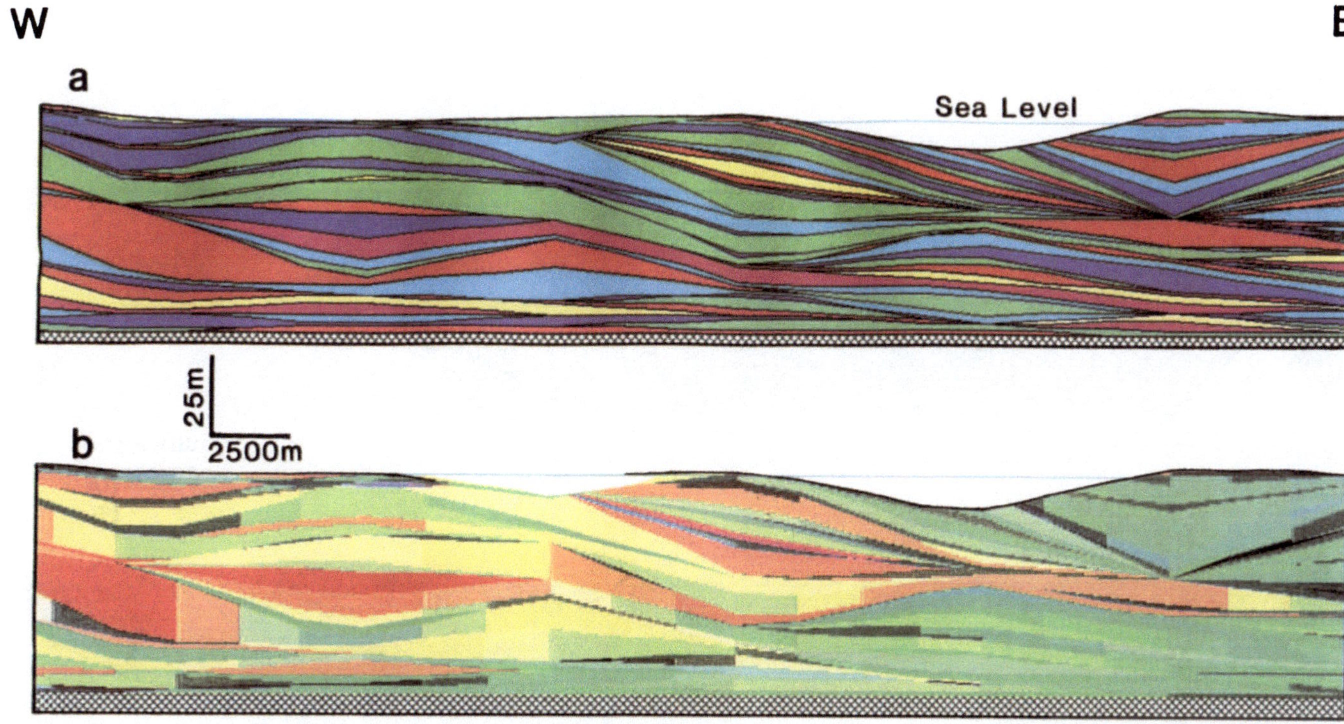

Plate 6 Enlargement of part of east-west section B–B′ showing deposits after 10,000 years simulated time. (*a*) Ages of beds. (*b*) Composition of beds.

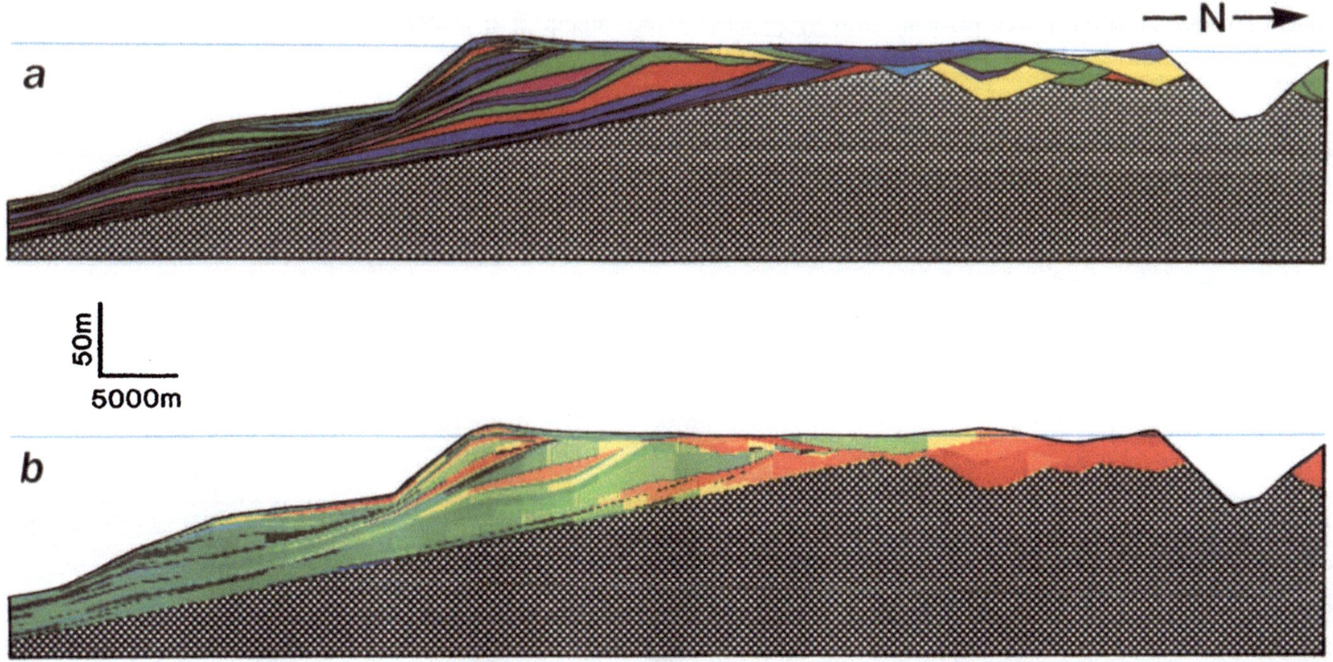

Plate 7 North-south section A–A′ through delta complex produced after 10,000 simulated years in experiment involving lower flow rate and lower proportion of coarse sediments than experiment shown in Plates 2, 3, 4, and 6. (*a*) Ages of beds. (*b*) Composition of beds.

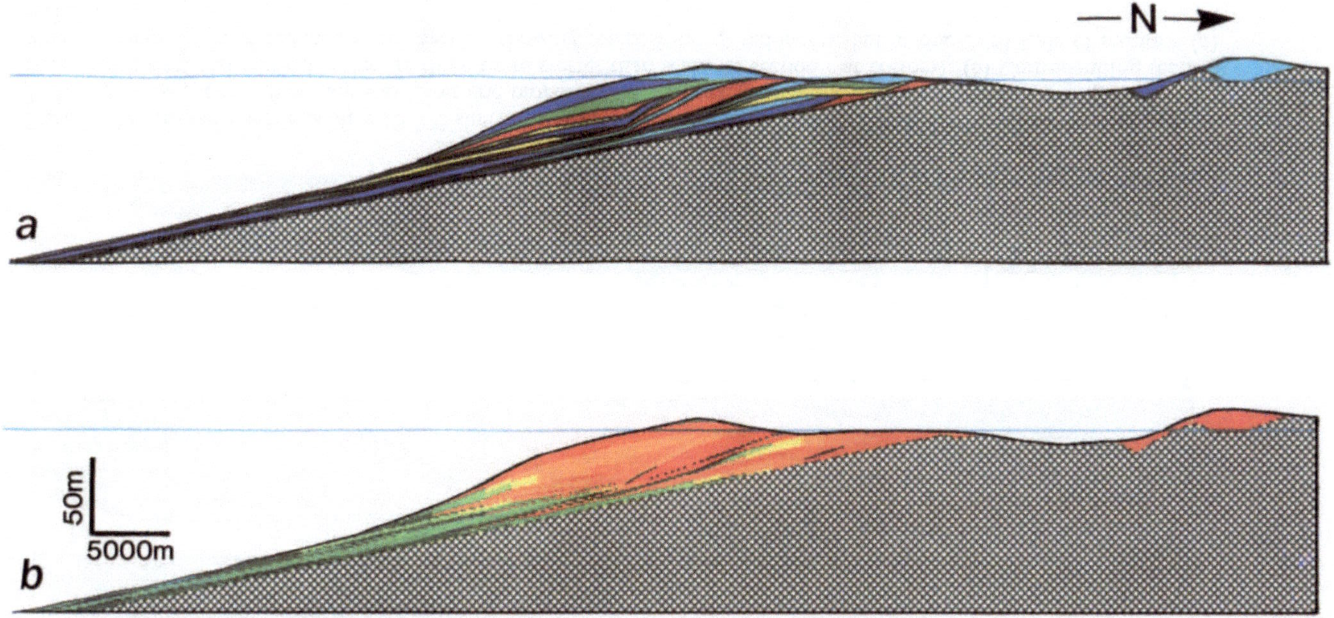

Plate 8 Sequence similar to that of Plate 7 except that experiment involved higher proportion of sand and higher flow rate than experiment that produced sequence shown in Plates 2, 3, 4, and 6. (*a*) Geologic age. (*b*) Composition.

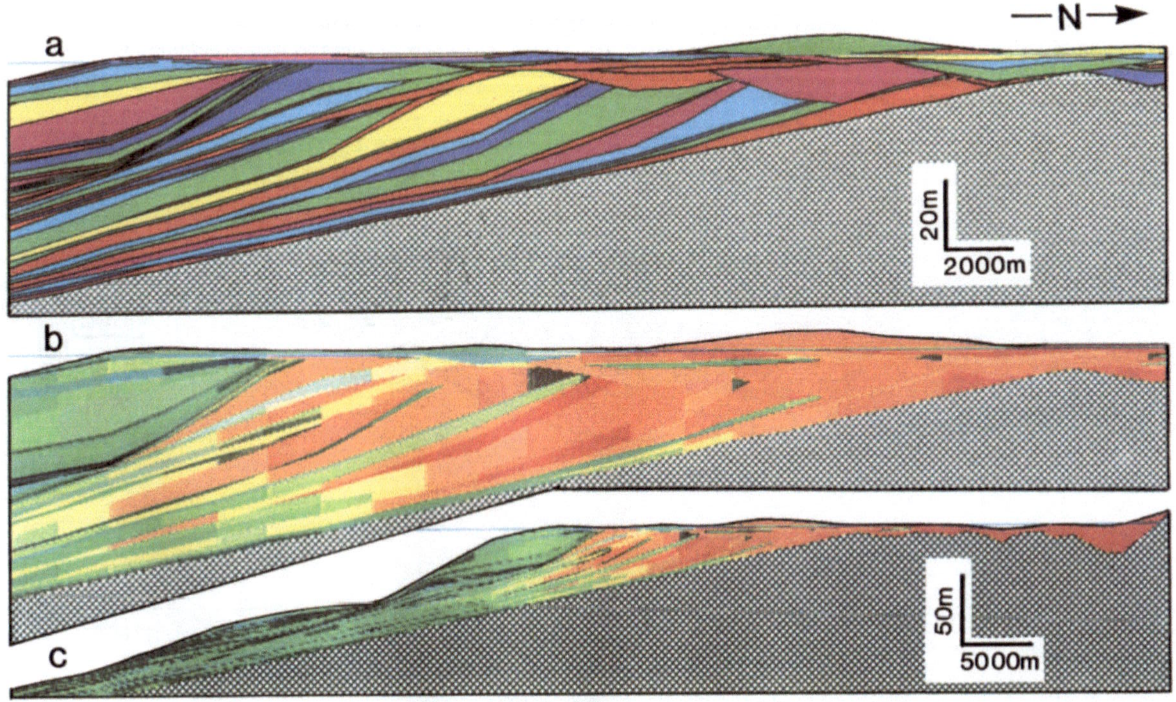

Plate 9 North-south sections after 10,000 simulated years produced under conditions similar to those shown in Plates 2, 3, 4, and 6, except that flow rate was lower and proportion of sand was greater. (*a*) Enlarged section showing ages of deposits (horizontal and vertical scales are three times greater than scales of section *c* at bottom). (*b*) Corresponding section showing composition of sediments. (*c*) Section showing composition of sediments, but at one-third scale of sections (*a*) and (*b*).

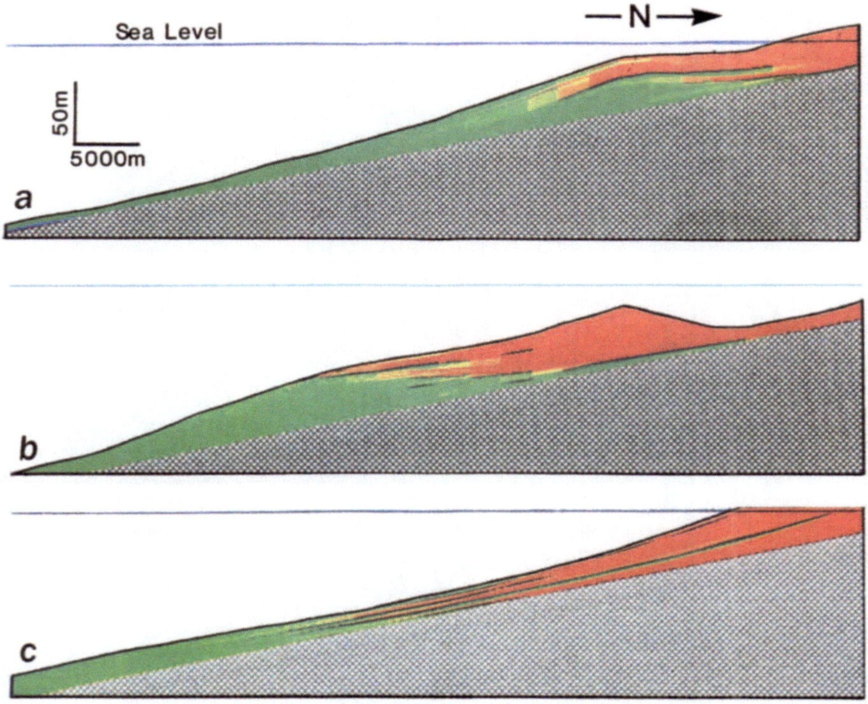

Plate 10 Cross sections produced in Golden Meadow experiments that are perpendicular to shoreline showing responses after 50,000 simulated years to three sets of hydraulic conditions. Proportion of sediment supplied by river has remained constant at 60% sand and 40% clay. (*a*) Response when 5000 m^3/s of water and 1570 kg/s of sediment were supplied. Simulated deposits contain more fine sediment than actual sequence observed in well logs. (*b*) Response when 7500 m^3/s of water and 5000 kg/s of sediment were supplied. Sand beds wedge out too rapidly from shore. (*c*) Response when 10,000 m^3/s of water and 5000 kg/s of sediment were supplied, providing "best approximation" of Golden Meadow deposits.

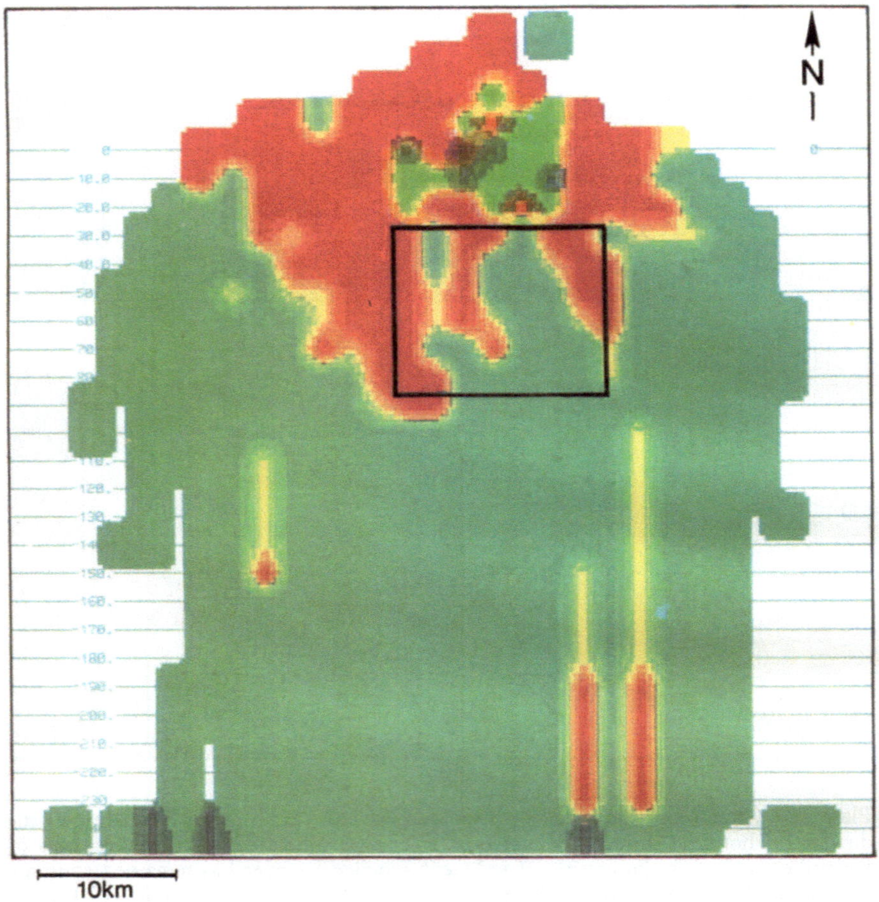

Plate 11 Regional facies map obtained in Golden Meadow experiment in which 10,000 m^3/s of water and 5000 kg/s of sediment were supplied for 50,000 simulated years. Sediment supplied by river consisted of 60% sand and 40% clay. Map shows proportion of sands in uppermost 30 m of deposited sediment. Red denotes sediment containing over 80% sand, whereas green indicates less than 20% sand. Orange, yellow, and yellow-green denote intermediate proportions of sand. Box outlines area enlarged in Plate 12, which contains Golden Meadow field.

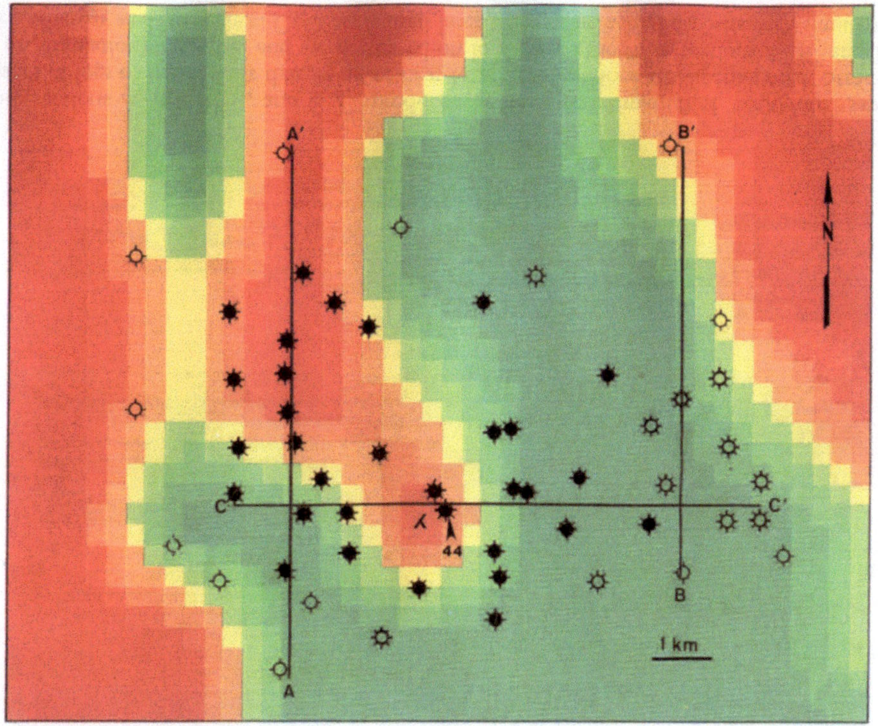

Plate 12 Enlarged map of area containing Golden Meadow field showing simulated facies. Location of area shown in Plate 11. X denotes location within lobate sand body that has been "predicted" to contain maximum thickness of sand. Traces of sections AA′, BB′, and CC′ (Plate 13), are shown.

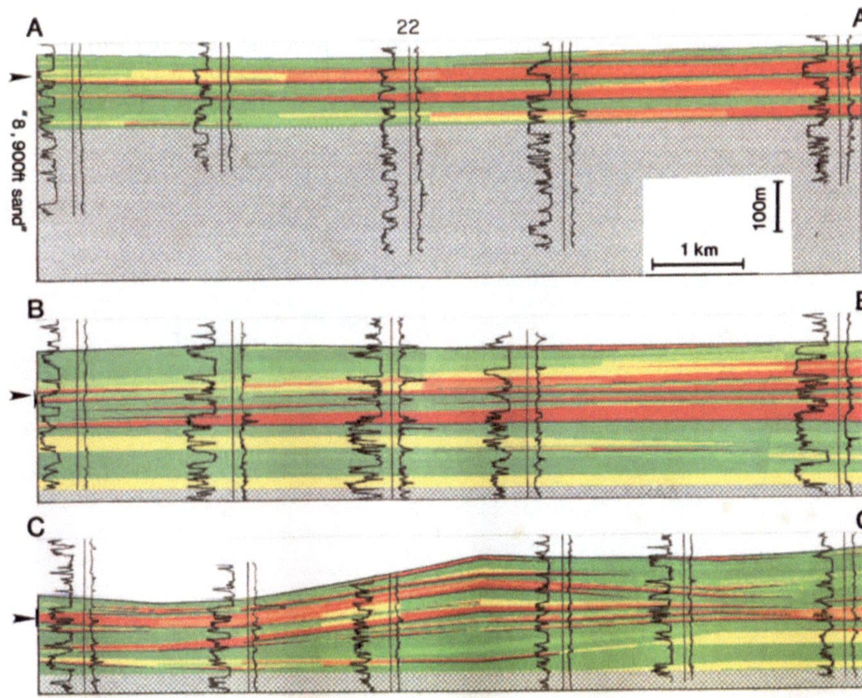

Plate 13 Stratigraphic sections A–A′, B–B′, and C–C′ produced by best approximation of "8900-foot" sand. Locations of sections are shown in Plate 12. Traces of logs of actual wells that lie on or near location of sections are superimposed. Stratigraphic correlations between wells are based solely on simulation's response, and are not based on logs. Wells shown in sections A–A′ and B–B′ were used later to calibrate simulation model, whereas wells in section C–C′ were used later to compare actual deposits with simulated deposits.

dips southward, and includes shales deposited offshore and interbedded sands and shales deposited closer to shore. Miocene and Pliocene sands produce at Golden Meadow.

We selected Golden Meadow field because it is structurally simple, in contrast with many other fields in the region, and because abundant information is provided by wells and was available to us. We used information from 55 wells, all deeper than 9000 feet. Some faults intersect the fields, but it is possible to make stratigraphic correlations across faults because persistent sands, such as the "8900-foot sand" (Fig. 9.9), are present and are easily recognized on the well logs. The 8900-foot sand occurs at depths of about 8900 feet and, though readily correlated from well to well, is highly variable laterally. It does not yield oil or gas, but is similar to deeper sands that produce at Golden Meadow. The 8900-foot sand consists of an upper sand and a lower sand separated by shale (Fig. 9.9). Both sands thin southward, away from the shoreline of the ancient Gulf of Mexico.

Our simulation involved a stratigraphic interval that extends from about 200 feet below the 8900-foot sand, to about 100 feet above. Our main objective has been to predict the properties of the sands over the area of the field. Although SEDSIM3 can handle four types of clastic sediment, we assumed that only sand and clay were present, increasing computing efficiency and simplifying interpretation. In the experiments, we repeatedly adjusted SEDSIM3's parameters to bring SEDSIM3's responses into accord with information provided by logs of only some of the available wells. We then compared responses with interpretations based on all of the available wells. For example, information from six wells was used to adjust SEDSIM3's input parameters. The remaining wells were then used to "verify" the degree to which the "predictions" made by SEDSIM3 were borne out.

The overall area of the simulation experiment is square, 65 km on a side, or roughly seven times the length and width of Golden Meadow field. We used a large area because it provided sufficient expanse so that we could look within it to find the subarea that seems to best approximate the depositional site of the 8900-foot sand in the Golden Meadow field. In other words, we needed enough elbow room so that we could best establish Golden Meadow's "location" within the overall simulated area. Furthermore, we wanted to ensure that the simulated deposits of interest formed far enough from the edges of the simulated area to avoid boundary problems.

In the experiment we assumed that the initial topography was similar to that of the present Gulf of Mexico, involving a gentle slope onshore and a slightly steeper slope (5 m/km) offshore. We placed a steady source of fluid and sediment representing the ancient Mississippi River onshore, and then allowed SEDSIM3 to run for 50,000 years of simulated time. We assumed that sand-clay ratios and ratios of sediment discharge relative to water discharge were about the same as those of the modern Mississippi River (Table 9.1). The principal challenge then was to choose an appropriate flow rate, because SED-

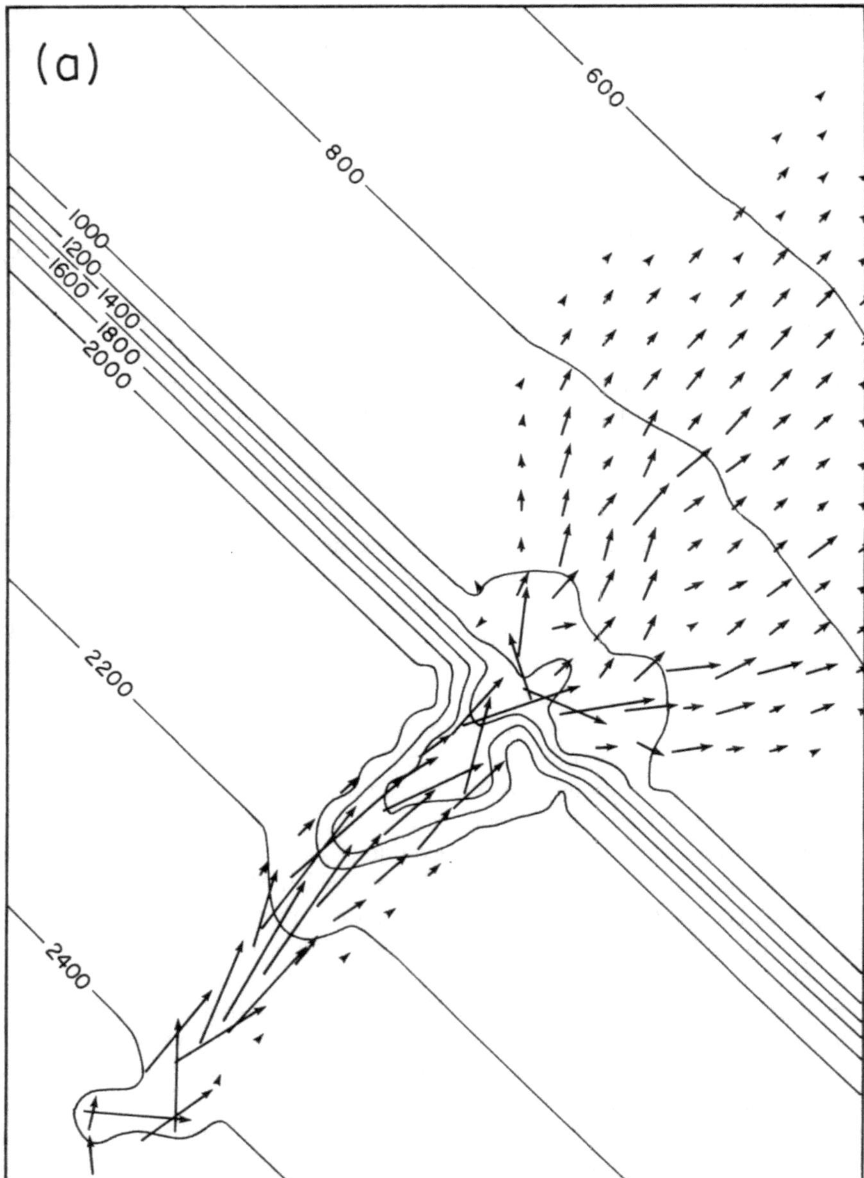

Figure 9.6 Maps showing submarine topography and flow vectors during simulation of Simpson Canyon. Contours in meters with respect to datum that is 2500 m below sea level. (*a*) Topography 20,000 years after canyon carving began. (*b*) Topography at 60,000 years.

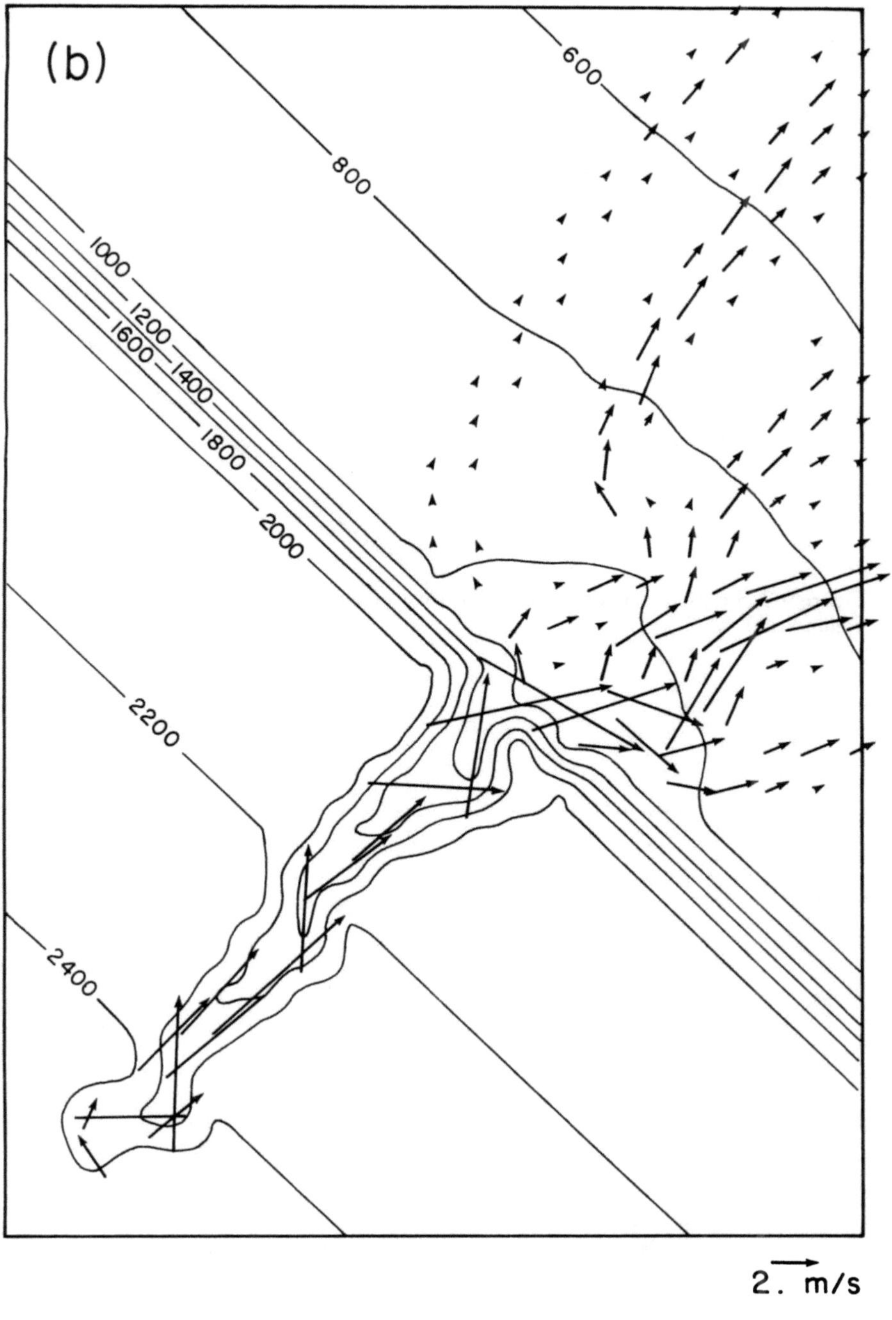

Figure 9.6 (continued)

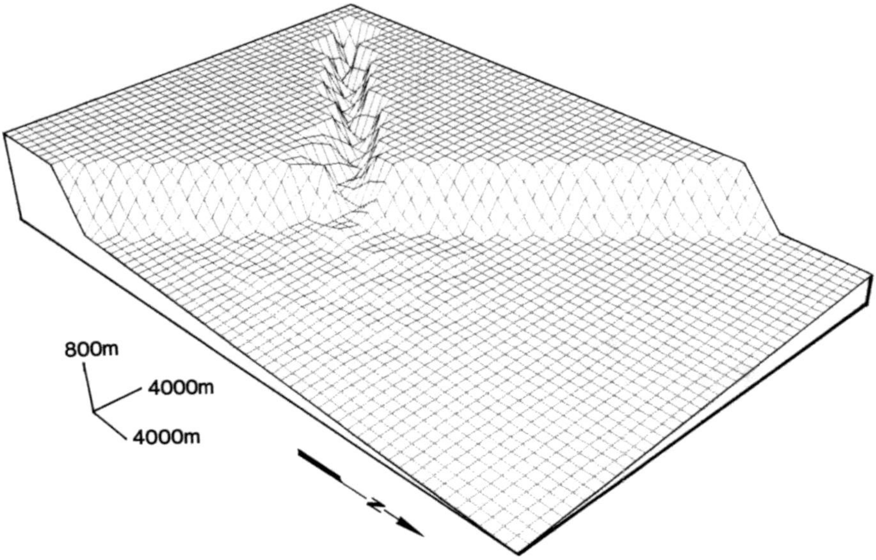

Figure 9.7 Simpson Canyon after 60,000 years of simulated time (corresponds to Figure 9.6*b*).

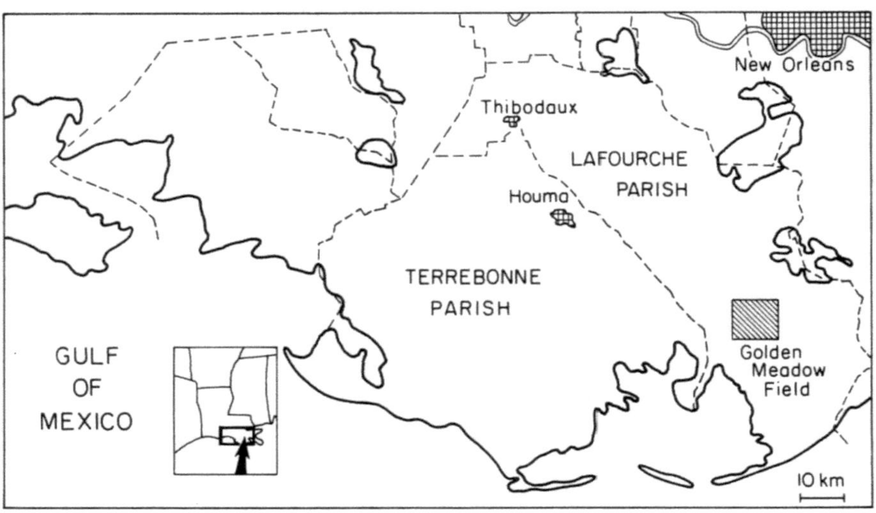

Figure 9.8 Index map of Golden Meadow oil and gas field.

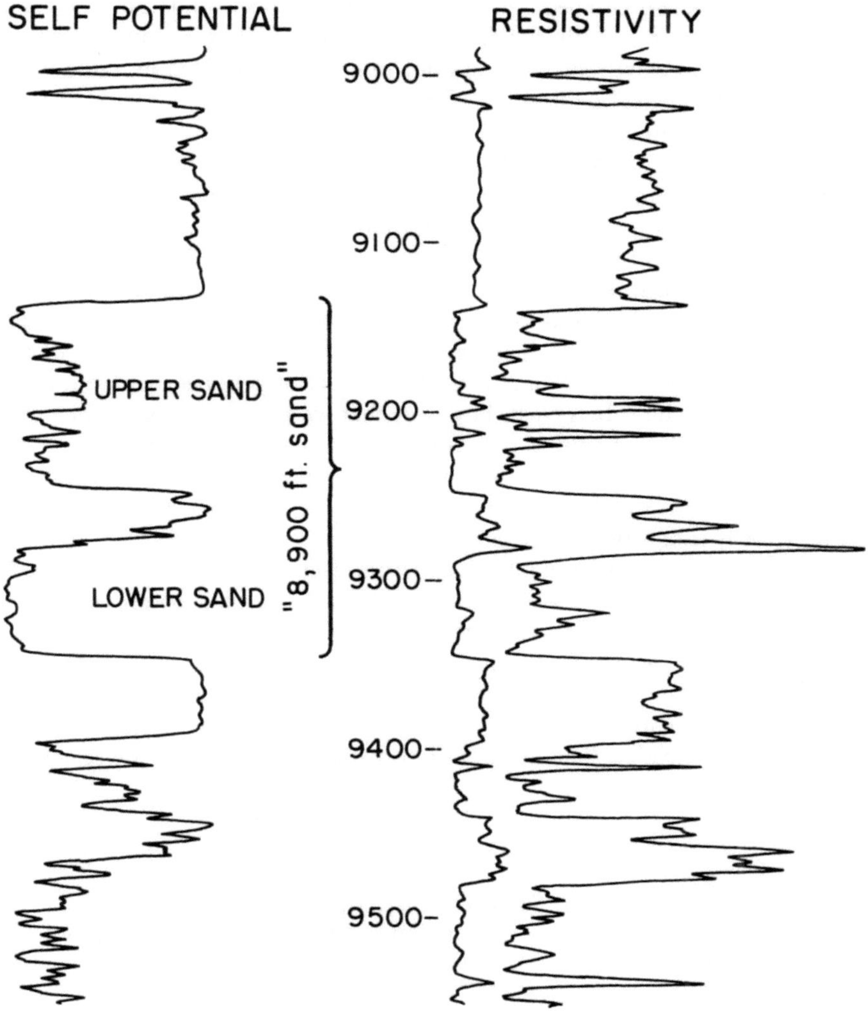

Figure 9.9 Segment of petrophysical log of well "22" (Plate 13 A-A′) in Golden Meadow field, showing "upper" and "lower" sand units that form "8900-foot" sand.

SIM3 is very sensitive to flow rates. We used data for the modern Mississippi River (Trowbridge, 1930; Saxena, 1976; Perlmutter, 1987), experimenting with various flow rates before we achieved a reasonable match between simulated and actual deposits.

The first experiment (Plate 10*a*) involves a discharge of 5000 m^3/s of water, about one third of the Mississippi's present discharge at normal stage, of about

15,000 m^3/s. We also assumed a volumetric sediment concentration of 0.00025, which corresponds to a sediment discharge of about 1570 kg/s, with sediment consisting of 40% clay, (corresponding to "fine" sediment listed in Table 9.1) and 60% sand ("coarse" sediment in Table 9.1).

The early experiments yielded beds that contain too much clay when compared with those of Golden Meadow, and we concluded that the flow rate was too low. Then we progressively increased the flow to 50,000 m^3/s. We also increased the aggregate sediment concentration to 0.0018, which represents a sediment discharge of about 111,000 kg/s, with a flow of 50,000 m^3/s. The higher values for both flow rate and sediment concentration yield deposits that accord better with the 8900-foot sand. Perhaps the "ancestral" Mississippi River had about the same discharge rates and sediment concentrations in late Miocene and early Pliocene time as it does today, or perhaps they were substantially greater then. It is clear, however, that we need more experience before SEDSIM3 will be useful in reconstructing the past status of rivers such as the ancestral Mississippi River. Of course, there were lesser rivers that flowed into the Gulf of Mexico in the past, just as there are today, so our experiments are necessarily simplistic and incomplete.

We performed ten experiments in all. After each experiment we compared the results with sections that involve detailed stratigraphic correlations between well logs. We tried to select the simulated sequence that best matches the actual sequence, which involves subjective comparisons. We compared proportions of sand and clay, thickness and shapes of individual sand and clay bodies, and areal variations in sand bodies. Furthermore, we shifted the location of Golden Meadow field about within the simulation area as we searched for the best match.

TABLE 9.1 Comparison of Sediment Discharge Rates for Various Particle Sizes with Respect to Various Water Discharge Rates in Modern Mississippi Delta

Water Discharge (m^3/s)	*Sediment Discharge ($\times 10^3$ kg/s)*				
	Coarse			*Fine*	
	Gravel	*Sand*	*Coarse Silt*	*Silt & Clay*	*Total*
	2000 μm	63 μm	37.5 μm		
5,000	0.00	0.47	0.47	0.63	1.57
15,000	0.00	4.75	1.58	12.78	19.02
50,000	3.17	53.90	22.20	31.71	111.0
100,000	6.34	95.13	79.27	72.93	253.7

Source: Data provided by Trowbridge (1930), Saxena (1976), and Perlmutter (1987).

An improved match is shown in Plate 10*b*, which involves a flow rate of 7500 m^3/s and a total sediment concentration of 0.0005 (yielding a sediment discharge of 5000 kg/s). Although the proportion of sand and clay in the simulated deposits is a good match, the simulated sand bodies are too limited laterally in the direction of progradation. When we increased the flow rate to 10,000 m^3/s and left the sediment input rate unchanged (thus reducing sediment concentration to 0.00037), the coarse sediment was carried farther out in the basin, yielding a still better match as shown in Plate 10*c*.

The sections and maps yielded by the "best matching" experiment over the range of conditions tested are shown in Plates 11 through 13. Comparisons show that the simulated beds are generally thinner than the actual beds, but simply doubling the vertical scale of the simulated deposits improves the match. This discrepancy may arise because SEDSIM3 does not incorporate isostatic compensation and compaction, both of which should allow individual beds to receive increased amounts of sediment.

Figure 9.10 is a topographic contour map of the best matching deltaic deposits after 50,000 simulated years. It suggests that the shoreline prograded approximately 30 km during this interval. Several lobate bodies of sand have been formed (Plates 11 and 12). Vertical sections through the simulated deposits provide comparison with the actual well logs, which are superimposed (Plate 13).

The simulated deposits obtained can be "explored" by producing sections and facies maps. Facies maps in Plates 11 and 12 display proportions of sand and clay within an interval 30 m thick that lies immediately below the top of the 8900-foot sand. The maps show the apex of a sand lens (indicated by X in Plate 12) that is not directly indicated by actual well logs, although the east-west section through the southern part of the simulated deposits (C-C′ in Plate 13) suggests that the sand thickens in the south-central part of the Golden Meadow field. The log of an actual well located near where SEDSIM3 "predicts" locally thicker sands (well 44 in Plate 12), may partly substantiate SEDSIM3's prediction.

ASSESSING THE GOLDEN MEADOW EXPERIMENTS

A sedimentary process simulation model such as SEDSIM3 provides an opportunity to explore many alternative assumptions. We can shift geographic locations in search for accord, we can adjust the discharges of rivers, or we can adjust sediment concentrations and proportions of grain sizes. No unique combination of variables may prove to be "best," but the experiments do permit us to explore the consequences when assumptions are changed.

What assumptions seem best at Golden Meadow? Tentatively, our experiments suggest that for an ancient Mississippi River, a fluid discharge rate of about 10,000 m^3/s and a sediment discharge rate of about 5000 kg/s are rea-

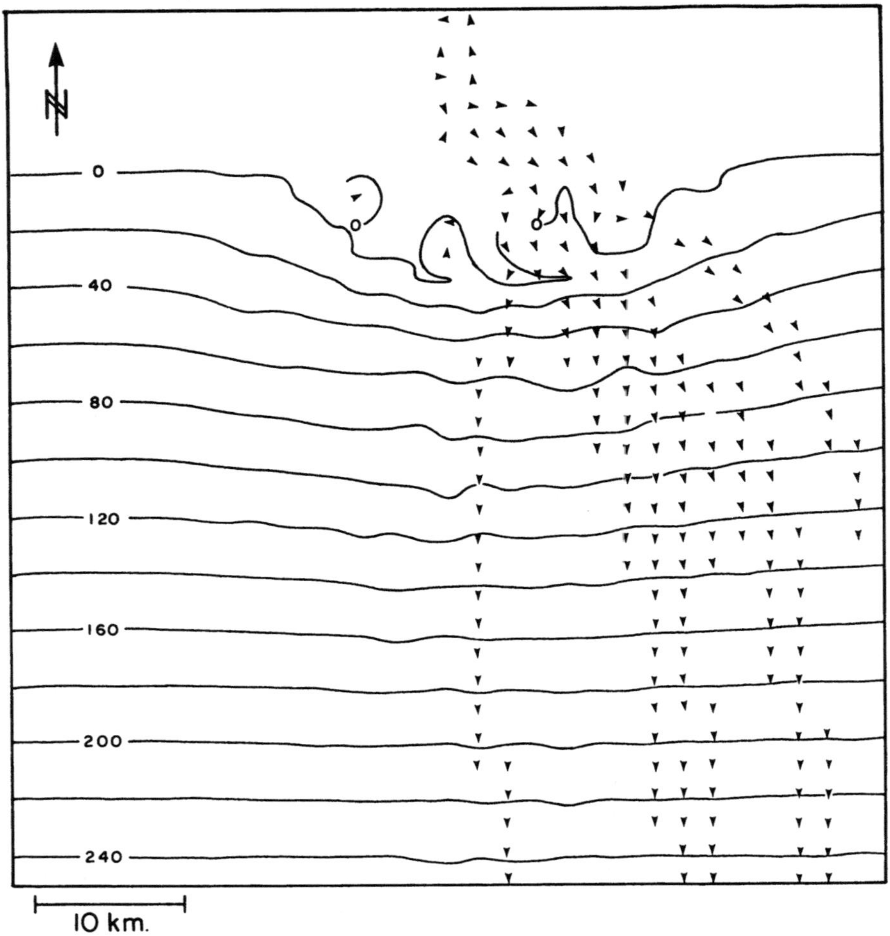

Figure 9.10 Map showing submerged topography of deltaic deposit produced when 10,000 m^3/s of water and 5000 kg/s of sediment were supplied for 50,000 years simulated time.

sonable, although both are less than present rates of the Mississippi River. Much more work would be needed before these estimates could be regarded as reliable. At best, though, estimates of ancient flow and sediment discharge rates are difficult to verify and are of only indirect use in making "spatial predictions" of sedimentary facies. On the other hand, SEDSIM3's abilities in predicting the geometrical form of bodies of sand and silt may be useful in oil and gas exploration. We invite our readers to adapt SEDSIM3 to their own research.

SUMMARY

SEDSIM3 has been applied to two real-world geologic settings, one in Simpson Canyon in northern Alaska and the other in Golden Meadow field in southern Louisiana. In both cases SEDSIM3 was able to reproduce actual sedimentary features observed either in seismic sections or in well logs. SEDSIM3 also provided predictions about the deposits in the general vicinity of seismic sections or wells. At Simpson Canyon, the predictions could not be verified because the critical seismic data are not available. At Golden Meadow, however, well data are abundant and permitted experiments involving repeated comparisons and adjustments.

APPENDIX A

Notation Conventions

PHYSICAL QUANTITIES

The following is a list of physical quantities used in this book. The list indicates the quantity by name, the symbol that represents it in the text, and the corresponding units in the International System (SI).

Quantity	*Symbol in text*	*Units (SI)*
Area	A	m^2
Acceleration	$\mathbf{a}$	m/s^2
Coefficient of bottom friction	c_1	dimensionless
Coefficient of lateral friction	c_2	kg/(m s)
Coefficient of transport	c_t	$m\ s^3/kg$
Coefficient of sediment transportability	f_1 or f_{1,K_s}	dimensionless
Coefficient of erosion-deposition	f_2 or f_{2,K_s}	m/s
Coefficient of movement threshold	f_3 or f_{3,K_s}	$kg/(m\ s^2)$
Coefficients of basement decomposition	f_{4,K_s}	dimensionless

Coefficient of transportability to fall velocity	c_f	s/m
Coordinates, spatial	x, y, z	m, m, m
Density	ρ	kg/m^3
Density, relative	R_ρ	dimensionless
Depth, relative	s	dimensionless
Energy	E	$kg\ m^2/s^2$
Flow depth	h	m
Flow rate per unit area	F_{s_i}	m/s
Flow surface elevation	H	m
Flow velocity (3 dimensions)	$\mathbf{q}$	m/s, m/s, m/s
Flow velocity components	u, v, w	m/s, m/s, m/s
Flow velocity, relative	r	dimensionless
Flow velocity (2 dimensions)	$\mathbf{Q}$	m/s, m/s
Fluid density	ρ	kg/m^3
Froude number	F_n	dimensionless
Gravity	$\mathbf{g}$	m/s^2
Hydraulic radius	R_h	m
Manning's roughness	n	$s/m^{1/3}$
Mass	m	kg
Momentum	$\mathbf{M}$	kg m/s
Number of columns, current column number	M, j	dimensionless
Number of rows, current row number	N, i	dimensionless
Number of elements, current element number	L, k	dimensionless
Number of sed. types, current type	N_s, K_s	dimensionless
Particle Reynolds number	R_*	dimensionless
Position (2 dimensions)	$\mathbf{X}$	m, m
Power	P	$kg\ m^2/s^3$
Pressure	p	$kg/(m\ s^2)$
Reynolds number	R_n	dimensionless
Sediment concentration	l	dimensionless
Sediment concentration, effective	Λ_e, Λ_{em}	dimensionless
Sediment particle diameter	d	m
Sediment particle fall velocity	W	m/s
Sediment transport capacity	Λ	dimensionless
Sediment transport rate	Q_s	kg/(m s)
Sediment type index	$K_s, \boldsymbol{\kappa}$	dimensionless
Shear stress	τ	$kg/(m\ s^2)$
Shields parameter	F_*	dimensionless
Slope of flow surface	S	dimensionless

Specific weight	γ	kg/(m^2 s^2)
Specific weight, relative	r_γ	dimensionless
Time	t	s
Topographic elevation	Z	m
Velocity, shear	v_*	m/s
Viscosity	μ	kg/(m s)
Viscosity, kinematic	ν	m^2/s
Volume	V	m^3

MATHEMATICAL OPERATORS

The following list explains the notation utilized throughout this book to indicate mathematical operators.

$$\nabla f = \frac{\partial f}{\partial x}, \frac{\partial f}{\partial y}, \frac{\partial f}{\partial z}$$

$$\nabla \mathbf{q} = \frac{\partial \mathbf{q}}{\partial x} + \frac{\partial \mathbf{q}}{\partial y} + \frac{\partial \mathbf{q}}{\partial z}$$

$$\nabla^2 f = \frac{\partial^2 f}{\partial x^2} + \frac{\partial^2 f}{\partial y^2} + \frac{\partial^2 f}{\partial z^2}$$

$$\nabla^2 \mathbf{q} = \frac{\partial^2 \mathbf{q}}{\partial x^2} + \frac{\partial^2 \mathbf{q}}{\partial y^2} + \frac{\partial^2 \mathbf{q}}{\partial z^2}$$

$$\nabla \cdot \mathbf{q} = \frac{\partial u}{\partial x} + \frac{\partial v}{\partial y} + \frac{\partial w}{\partial z} \qquad \mathbf{q} = (u, v, w)$$

$$(\mathbf{q} \cdot \nabla)\mathbf{r} = u\frac{\partial \mathbf{r}}{\partial x} + v\frac{\partial \mathbf{r}}{\partial y} + w\frac{\partial \mathbf{r}}{\partial z}$$

$$\frac{D\mathbf{q}}{Dt} = \frac{\partial \mathbf{q}}{\partial t} + (\mathbf{q} \cdot \nabla)\mathbf{q}$$

MATHEMATICAL SYMBOLS

d = derivative
∂ = partial derivative
D = derivative following fluid motion
$\int$ = integral
$\oint$ = curvilinear integral
$\cdot$ = scalar product
Σ = summation
$|\,|$ = absolute value
Δ = finite increment

APPENDIX B

Listing of Program SEDCYC3

Program SEDCYC3 simulates flow, erosion, and deposition with up to four sediment types. A complete listing of the main program, plus the subroutines in alphabetical order, is provided. To generate a data file for SEDCYC3, the user would build a short main program that calls subroutine WRITDF. Parameters then can be assigned inside the main program, or the file written by WRITDF can be edited so as to write in the parameters.

```
C.....PROGRAM SEDCYC3
      IMPLICIT COMPLEX (C)
      CHARACTER*72 IT
      COMMON /BLK1/LBIT(50,41,61)
      COMMON IT,TO,TR,TD,DT,TB,TE,TID,TF,FLD,SWD,VIS,ROU,DV,CVC,CWV,
     & SDI(4),SDD(4),BD(4),FCO(5),
     & TI,LI,CPI(20),CVI(20),SLI(20,4),TEV,LEV,CPEV(20),
     & DX,N,M,NQ,QX(10,4),QY(10,4),QZ(10,4),
     & ZT(41,61),ZB(41,61),DZZ,L,CP(5000),CV(5000),SL(5000,4),
     & FEN(41,61),CVN(41,61),SUB(41,61),ZT1(41,61),SDER(41,61),
     & NY,G,DX2,DZ,DM,FF,FB,F3,F4,VST(4),FU(4),FD(4),TC(4),FAC
      F1=3.1E+07
      E=1.00001
C.....READ
      OPEN (20,FILE='for20.d',FORM='FORMATTED')
      REWIND (20)
      CALL READDF3
      CLOSE (20)
C.....OPEN MESSAGE FILE
      OPEN (25,FILE='for25.d',FORM='FORMATTED')
      REWIND (25)
      WRITE (25,1) TO,TR,L
    1 FORMAT (1X,' TIME=',F15.7,' YEARS OF ',F15.7,' YEARS'/
     & ' L=',I6)
      CLOSE (25)
```

```
      OPEN (22,FILE='for22.d',FORM='UNFORMATTED')
      REWIND (22)
      CALL FILTER3 (-1)
      IF (T0.NE.0.) GOTO 2
      CALL NEWNOD3
      CALL WRITGU3
      IF (TR-T0.LT.0.0000001) STOP
    2 T=T0
      T1=DT*TB/F1
      T2=(TE+TID)*TF
      NYET=INT((TR-T0)/T2-0.00001)
      NYT=TE*F1/DT/TB
      CALL INFLOW3
C.....ENTER MAIN LOOPS
      TYE=0.
      DO 100 NYE=0,NYET
      CALL SET3
      DO 10 NY=0,NYT
      TY=(NY+1)*T1
      CALL NEWNOD3
      CALL NEWELM3
      CALL FILTER3 (1)
      IF (TI.LE.TE.AND.AMOD(TY*E,TI).LT.T1.AND.TI.GT.0.) CALL INFLOW3
      T0=T+TYE+TY
      IF (TD.LE.TE.AND.AMOD(TY*E,TD).LT.T1.AND.TD.GT.0.) CALL WRITGU3
   10 CONTINUE
      TYE=(NYE+1)*T2
      T0=T+TYE
      IF (TF.GT.1.001) CALL EXTRAP3
      IF (TI.GT.TE.AND.AMOD(TYE*E,TI).LT.T2) CALL INFLOW3
      IF (TD.GT.TE.AND.AMOD(TYE*E,TD).LT.T2) CALL WRITGU3
C.....WRITE CURRENT TIME
      OPEN (25,FILE='for25.d',FORM='FORMATTED')
      REWIND (25)
      WRITE (25,1) T0,TR,L
      CLOSE (25)
  100 CONTINUE
C.....END
      OPEN (21,FILE='for21.d',FORM='FORMATTED')
      REWIND (21)
      CALL WRITDF3
      CLOSE (21)
      STOP
      END
C-----------------------------------------------------------------------
      SUBROUTINE EXTRAP3
C.....EXTRAPOLATES TIME SINCE LAST CALL SET BY FACTOR F (F>1.)
      IMPLICIT COMPLEX (C)
      DIMENSION LC(800)
      CHARACTER*72 IT
      COMMON /BLK1/LBIT(50,41,61)
      COMMON IT,T0,TR,TD,DT,TB,TE,TID,TF,FLD,SWD,VIS,ROU,DV,CVC,CWV,
     & SDI(4),SDD(4),BD(4),FCO(5),
     & TI,LI,CPI(20),CVI(20),SLI(20,4),TEV,LEV,CPEV(20),
     & DX,N,M,NQ,QX(10,4),QY(10,4),QZ(10,4),
     & ZT(41,61),ZB(41,61),DZZ,L,CP(5000),CV(5000),SL(5000,4),
     & FEN(41,61),CVN(41,61),SUB(41,61),ZT1(41,61),SDER(41,61),
     & NY,G,DX2,DZ,DM,FF,FB,F3,F4,VST(4),FU(4),FD(4),TC(4),FAC
C.....ENTER MAIN LOOPS
      DO 10 I=1,N
      DO 10 J=1,M
      ZTN=ZT(I,J)+(ZT(I,J)-ZT1(I,J))*TF
      IF (ZTN.LE.ZT(I,J)) GOTO 9
      K=(ZT(I,J)-ZB(I,J))/DZZ+0.99999
      K1=(ZT1(I,J)-ZB(I,J))/DZZ+0.99999
      IF (K1.LE.0) K1=1
      K2=(ZTN-ZB(I,J))/DZZ+0.99999
      DO 5 KK=K1,K
      K11=KK-K1+1
      LC(K11)=0
    5 CALL SEDNOD3 (I,J,KK,LC(K11))
      KK=0
      DO 6 KL=K+1,K2
```

```
      KK=KK+1
      IF (KK.GT.K) KK=1
    6 CALL SEDNOD3 (I,J,KL,LC(KK))
    9 ZT(I,J)=ZTN
      IF (ZTN.LT.ZB(I,J)) ZB(I,J)=ZTN
   10 CONTINUE
      RETURN
      END
C-----------------------------------------------------------------------
      SUBROUTINE FILTER3 (IFLAG)
C.....APPLIES TIME AND SPACE FILTERS TO VELOCITIES AND FLUID LEVELS
C     IFLAG=1 ADD VARIABLES
C     IFLAG=0 DIVIDE TO FIND AVERAGE
C     IFLAG=-1 INITIALIZE TO 0
      IMPLICIT COMPLEX (C)
      CHARACTER*72 IT
      COMMON /BLK1/LBIT(50,41,61)
      COMMON IT,T0,TR,TD,DT,TB,TE,TID,TF,FLD,SWD,VIS,ROU,DV,CVC,CWV,
     & SDI(4),SDD(4),BD(4),FCO(5),
     & TI,LI,CPI(20),CVI(20),SLI(20,4),TEV,LEV,CPEV(20),
     & DX,N,M,NQ,QX(10,4),QY(10,4),QZ(10,4),
     & ZT(41,61),ZB(41,61),DZZ,L,CP(5000),CV(5000),SL(5000,4),
     & FEN(41,61),CVN(41,61),SUB(41,61),ZT1(41,61),SDER(41,61),
     & NY,G,DX2,DZ,DM,FF,FB,F3,F4,VST(4),FU(4),FD(4),TC(4),FAC
      COMMON /FIL/CV1(41,61),FE1(41,61),NFT
      IF (IFLAG.EQ.0) GOTO 5
      IF (IFLAG.EQ.-1) GOTO 2
      NFT=NFT+1
C.....ADD CURRENT GRID TO PREVIOUS ONE
      DO 3 I=1,N
      DO 3 J=1,M
      CV1(I,J)=CV1(I,J)+CVN(I,J)
    3 FE1(I,J)=FE1(I,J)+FEN(I,J)
      RETURN
C.....FIND TIME AVERAGES
    5 IF (NFT.EQ.0) RETURN
      DO 4 I=1,N
      DO 4 J=1,M
      CVN(I,J)=CV1(I,J)/NFT
    4 FEN(I,J)=FE1(I,J)/NFT
C.....FIND SPATIAL AVERAGES
      IF (NFS.LT.1) RETURN
      DO 11 IFS=1,NFS
      DO 10 I=1,N
      DO 10 J=1,M
      CV2=(0.,0.)
      FE2=0.
      NS=0
      IF (I.LE.1) GOTO 6
      CV2=CV2+CVN(I-1,J)
      FE2=FE2+FEN(I-1,J)
      NS=NS+1
    6 IF (I.GE.N) GOTO 7
      CV2=CV2+CVN(I+1,J)
      FE2=FE2+FEN(I+1,J)
      NS=NS+1
    7 IF (J.LE.1) GOTO 8
      CV2=CV2+CVN(I,J-1)
      FE2=FE2+FEN(I,J-1)
      NS=NS+1
    8 IF (J.GE.M) GOTO 9
      CV2=CV2+CVN(I,J+1)
      FE2=FE2+FEN(I,J+1)
      NS=NS+1
    9 CV1(I,J)=(CVN(I,J)*2.+CV2)/(2+NS)
   10 FE1(I,J)=(FEN(I,J)*2.+FE2)/(2+NS)
C.....EQUALIZE
      DO 11 I=1,N
      DO 11 J=1,M
      CVN(I,J)=CV1(I,J)
   11 FEN(I,J)=FE1(I,J)
C.....INITIALIZE TO 0
    2 DO 1 I=1,N
```

```
      DO 1 J=1,M
      CV1(I,J)=(0.,0.)
    1 FE1(I,J)=0.
      NFT=0
      RETURN
      END
C-----------------------------------------------------------------------
      SUBROUTINE HSLOP3(CPOS,ZE,CS,I,J,ICALL1,IFL)
C.....CALCULATES HEIGHT AND SLOPE AT A GIVEN POSITION
      IMPLICIT COMPLEX (C)
      CHARACTER*72 IT
      COMMON IT,T0,TR,TD,DT,TB,TE,TID,TF,FLD,SWD,VIS,ROU,DV,CVC,CWV,
     & SDI(4),SDD(4),BD(4),FCO(5),
     & TI,LI,CPI(20),CVI(20),SLI(20,4),TEV,LEV,CPEV(20),
     & DX,N,M,NQ,QX(10,4),QY(10,4),QZ(10,4),
     & ZT(41,61),ZB(41,61),DZZ,L,CP(5000),CV(5000),SL(5000,4),
     & FEN(41,61),CVN(41,61),SUB(41,61),ZT1(41,61),SDER(41,61),
     & NY,G,DX2,DZ,DM,FF,FB,F3,F4,VST(4),FU(4),FD(4),TC(4),FAC
C.....SET PARAMETERS
      R1=(FLD-SWD)/SWD
      R2=(SWD-FLD)/FLD
      IFL=0
C.....FIND CELL AND CHECK FOR OUT OF BOUNDS
      JC=REAL(CPOS)/DX+1
      IF (JC.LT.1) RETURN
      IC=N-AIMAG(CPOS)/DX
      IF (JC.LT.M.AND.IC.GE.1.AND.IC.LT.N) GOTO 7
      CPOS=CMPLX(-DX,-DX)
      RETURN
C.....FIND NODE
    7 J=REAL(CPOS)/DX+1.5
      I=N-AIMAG(CPOS)/DX+0.5
      ICALL=ICALL1
      I1=I
      J1=J
C.....FIND RELATIVE POSITION IN CELL
      PCX=REAL(CPOS)/DX-JC+1.
      PCY=AIMAG(CPOS)/DX-N+IC+1.
C.....SET PARAMETERS FOR SURFACE GEOMETRY
      A2=ZT(IC,JC)
      A3=ZT(IC,JC+1)
      A4=ZT(IC+1,JC+1)
      A1=ZT(IC+1,JC)
      ZZ=A2+A4-A1-A3
      ZEA=A1+(A4-A1)*PCX+(A2-A1)*PCY-ZZ*PCX*PCY
      CSA=CMPLX(A1-A4+ZZ*PCY,A1-A2+ZZ*PCX)/DX
      IF (ICALL.EQ.2) GOTO 2
      FEN(I,J)=FEN(I,J)-1.
    2 B2=FEN(IC,JC)
      B3=FEN(IC,JC+1)
      B4=FEN(IC+1,JC+1)
      B1=FEN(IC+1,JC)
      FEN(I1,J1)=FEN(I1,J1)+1.
      ZZ=B2+B4-B1-B3
      ZEB=(B1+(B4-B1)*PCX+(B2-B1)*PCY-ZZ*PCX*PCY)*DZ
      CSB=CMPLX(B1-B4+ZZ*PCY,B1-B2+ZZ*PCX)/DX*DZ
      ZE=ZEA+ZEB
      CS=CSA+CSB
      IF (SWD.LE.FLD) GOTO 10
      IF (ZEB*R2.LT.ZE) RETURN
      ZE=ZEB*R2
      CS=CSB*R2
      IFL=1
      RETURN
   10 IF (ZEA.GT.-ZEB) RETURN
      ZE=ZE*R1
      CS=CS*R1
      RETURN
      END
C-----------------------------------------------------------------------
      SUBROUTINE INFLOW3
```

```
C.....ALLOCATES NEW ELEMENTS FLOWING INTO THE SYSTEM
      IMPLICIT COMPLEX (C)
      CHARACTER*72 IT
C.....FOR STANDARD FORTRAN CHANGE 'COMMON' TO 'COMMON',
      COMMON /BLK1/LBIT(50,41,61)
      COMMON IT,T0,TR,TD,DT,TB,TE,TID,TF,FLD,SWD,VIS,ROU,DV,CVC,CWV,
     & SDI(4),SDD(4),BD(4),FCO(5),
     & TI,LI,CPI(20),CVI(20),SLI(20,4),TEV,LEV,CPEV(20),
     & DX,N,M,NQ,QX(10,4),QY(10,4),QZ(10,4),
     & ZT(41,61),ZB(41,61),DZZ,L,CP(5000),CV(5000),SL(5000,4),
     & FEN(41,61),CVN(41,61),SUB(41,61),ZT1(41,61),SDER(41,61),
     & NY,G,DX2,DZ,DM,FF,FB,F3,F4,VST(4),FU(4),FD(4),TC(4),FAC
      KI=0
C.....USE ELEMENTS THAT ARE OUT OF BOUNDS
      DO 1 K=1,L
      IF (KI.GE.LI) RETURN
      IF (REAL(CP(K)).GE.0.) GOTO 1
      KI=KI+1
      CP(K)=CPI(KI)
      CV(K)=CVI(KI)
      DO 5 LIT=1,4
    5 SL(K,LIT)=SLI(KI,LIT)
      IF (KI.GE.LI) RETURN
    1 CONTINUE
      K=L
C.....ALLOCATE NEW ELEMENTS
      KI1=KI+1
      DO 2 KI=KI1,LI
      K=K+1
      IF (K.GT.5000) GOTO 3
      CP(K)=CPI(KI)
      CV(K)=CVI(KI)
      DO 2 LIT=1,4
    2 SL(K,LIT)=SLI(KI,LIT)
      L=K
      RETURN
    3 open (24,FILE='for24.d',FORM='FORMATTED')
      WRITE (24,4)
    4 FORMAT (' CANNOT ALLOCATE ANY MORE ELEMENTS')
      STOP
      END
C----------------------------------------------------------------------
      SUBROUTINE NEWELM3
C.....FINDS NEW POSITIONS AND VELOCITIES OF FLUID ELEMENTS
      IMPLICIT COMPLEX (C)
      CHARACTER*72 IT
      COMMON /BLK1/LBIT(50,41,61)
      COMMON IT,T0,TR,TD,DT,TB,TE,TID,TF,FLD,SWD,VIS,ROU,DV,CVC,CWV,
     & SDI(4),SDD(4),BD(4),FCO(5),
     & TI,LI,CPI(20),CVI(20),SLI(20,4),TEV,LEV,CPEV(20),
     & DX,N,M,NQ,QX(10,4),QY(10,4),QZ(10,4),
     & ZT(41,61),ZB(41,61),DZZ,L,CP(5000),CV(5000),SL(5000,4),
     & FEN(41,61),CVN(41,61),SUB(41,61),ZT1(41,61),SDER(41,61),
     & NY,G,DX2,DZ,DM,FF,FB,F3,F4,VST(4),FU(4),FD(4),TC(4),FAC
C.....DEFINE SCALAR PRODUCT
      SCAPRO(C1,C2)=REAL(C1)*REAL(C2)+AIMAG(C1)*AIMAG(C2)
C.....SET CONSTANTS
      BDT=1.-FCO(5)
C.....ENTER MAIN DO LOOP, ONCE PER EACH ELEMENT
      DO 7 K=1,L
C.....FIND HEIGHT AND SLOPE AND CHECK FOR OUT OF BOUNDS
      CP0=CP(K)
      CALL HSLOP3(CP0,ZE,CS,I,J,1,IFL)
      CP(K)=CP0
      IF (REAL(CP(K)).LT.0.) GOTO 7
C.....FIND RELATIVE FRICTION COEFFICIENTS
      SB=DX2/(FEN(I,J)+1.)
      SF=(DX2-SB)*DZ/DX
      CVE=CVN(I,J)*SF/(SB+SF)
      F2=VIS*(SF+ROU*SB)/(SF+SB)
      CVT=CV(K)-CVE
      SS=CABS(CS)
```

```
      SQ=SQRT(1.+SCAPRO(CS,CS))
      CVF=0.
      IF (SS.NE.0..AND.F2.NE.0.)
     & CVF=SQRT(DM*G/F2*SS/SQ)*CS/SS
      CVS=G*CS/SQ*DT
      IF (F2.NE.0.) CVT=CVF+(CVT+CVS-CVF)*EXP(-F2*CABS(CVT-CVF)*DT/DM)
      IF (F2.EQ.0.) CVT=CVT+CVS
      CVT=CVT+CVE
C.....FIND TRANSPORT CAPACITY DIFFERENCE TCD
      IF (I.LE.1.OR.I.GE.N.OR.J.LE.1.OR.J.GE.M) GOTO 9
      V=CABS(CVN(I,J))
      DDV=CABS(CVN(I,J))**2*FB*SB*DT/DM
      SLTX=SL(K,1)/TC(1)+SL(K,2)/TC(2)+SL(K,3)/TC(3)+
     & SL(K,4)/TC(4)
      TCMAX=(DV/TC(1)+DV/TC(2)+DV/TC(3)+DV/TC(4))/4./100.
      TCD=AMIN1((2.*V*DDV+DDV*DDV)/(2.*DT)*DM,TCMAX)*TB-SLTX
C.....UPDATE VELOCITY AND POSITION
    9 CVW=CVT
    2 CVM=(CV(K)+CVW)/2.
      CPW=CP(K)+CVM*DT/SQ
C.....ERODE OR DEPOSIT
      JC=REAL(CP(K))/DX+1.
      IC=N-AIMAG(CP(K))/DX
      II=0
      JJ=0
      IF (REAL(CV(K)).LT.0.) JJ=1
      IF (AIMAG(CV(K)).GT.0.) II=1
      J=JC+JJ
      I=IC+II
      IF (I.LE.1.OR.I.GE.N.OR.J.LE.1.OR.J.GE.M) GOTO 5
      ZD1=ZT(I,J)-ZB(I,J)
    6 IF (TCD.LT.-0.00001.AND.SDER(I,J).LT.0.001) GOTO 3
      IF (TCD.GT.0.00001.AND.IFL.NE.1.AND.SDER(I,J).GT.-0.001) GOTO 4
      GOTO 17
C.....ERODE (TCD POSITIVE)
    4 K1=ZD1/DZZ+0.9999
   14 IF (K1.LE.0) GOTO 30
      LIT=0
      CALL SEDNOD3 (I,J,K1,LIT)
      ZER=TCD*TC(LIT)/DX2*FU(LIT)
      IF (ZER.LT.0.) GOTO 17
      IF (INT((ZD1-ZER)/DZZ+0.9999).GE.K1) GOTO 20
C.....ERODE TO CELL BOTTOM
      DZREM=ZD1-(K1-1)*DZZ
      SL(K,LIT)=SL(K,LIT)+DZREM*DX2
      ZT(I,J)=ZT(I,J)-DZREM
      TCD=TCD-DZREM*DX2/TC(LIT)/FU(LIT)
      IF (TCD.LE.0.) GOTO 17
      ZD1=ZD1-DZREM
      K1=K1-1
      GOTO 14
C.....ERODE CELL FRACTION
   20 SL(K,LIT)=SL(K,LIT)+ZER*DX2
      ZT(I,J)=ZT(I,J)-ZER
      GOTO 17
C.....ERODE BASEMENT
   30 ZER=TCD*FAC*bdt
      DO 21 LIT=1,4
   21 SL(K,LIT)=SL(K,LIT)+ZER*DX2*BD(LIT)
      ZT(I,J)=ZT(I,J)-ZER
      ZB(I,J)=ZT(I,J)
      GOTO 17
C.....DEPOSIT (TCD NEGATIVE)
C.....ENTER LOOP ONCE PER LITHOLOGY
    3 DO 38 LT=1,4
      LIT=LT
      K1=(ZT(I,J)-ZB(I,J))/DZZ+1
      SDP=-TCD*TC(LIT)*FD(LIT)
      ZDP=SDP/DX2
      IF (SDP/FD(LIT).LT.SL(K,LIT)) GOTO 40
C.....DEPOSIT WHOLE LITHOLOGY
      IF (SL(K,LIT).LE.0.00001) GOTO 38
```

```
      ZT(I,J)=ZT(I,J)+SL(K,LIT)/DX2*FD(LIT)
      SL(K,LIT)=SL(K,LIT)*(1.-FD(LIT))
      K2=(ZT(I,J)-ZB(I,J))/DZZ+1
      DO 35 K3=K1,K2
   35 CALL SEDNOD3 (I,J,K3,LIT)
      TCD=TCD+SL(K,LIT)/TC(LIT)
   38 CONTINUE
      GOTO 17
C.....DEPOSIT LITH FRACTION
   40 K1=(ZT(I,J)-ZB(I,J))/DZZ+1
      SL(K,LIT)=SL(K,LIT)-SDP
      ZT(I,J)=ZT(I,J)+ZDP
      K2=(ZT(I,J)-ZB(I,J))/DZZ+1
      DO 45 K3=K1,K2
   45 CALL SEDNOD3 (I,J,K3,LIT)
   17 CONTINUE
    5 CV(K)=CVW
      CPO=CP(K)
      IF (CABS(CPO-CPW).LT.DX.OR.IERR1.EQ.1) GOTO 15
      OPEN (24,FILE='for24.d',FORM='FORMATTED')
      rewind (24)
      WRITE (24,8) TO,K,CPO,CPW,CVW,R,AB,VS,ZDIF,FEN(I,J)
    8 FORMAT (' WARNING: TIME INCREMENT 1 OR FACT 1 TOO HIGH'/
     &' TO,K,CPO,CPW,CVW,R,AB,VS,ZDIF,FEN(I,J)'/
     &1X,F8.5,I2,1X,6F6.0,2E10.2/1X,3F5.1)
      IERR1=1
      CLOSE (24)
   15 CP(K)=CPW
C.....COMMENT IN NEXT LINE MUST BE ELIMINATED TO PLOT REAL-TIME PATHS
C     CALL PLTPAT3 (CPO,K,IPLT,K)
      IF (ZEW.LT.0.) CP(K)=CP(K)+(CVC+CWV/(-ZEW+FEN(I,J)*DZ))*DT
    7 CONTINUE
      RETURN
      END
C-----------------------------------------------------------------------
      SUBROUTINE NEWNOD3
C.....UPDATES NODE PARAMETERS
      IMPLICIT COMPLEX (C)
      CHARACTER*72 IT
      COMMON /BLK1/LBIT(50,41,61)
      COMMON IT,TO,TR,TD,DT,TB,TE,TID,TF,FLD,SWD,VIS,ROU,DV,CVC,CWV,
     & SDI(4),SDD(4),BD(4),FCO(5),
     & TI,LI,CPI(20),CVI(20),SLI(20,4),TEV,LEV,CPEV(20),
     & DX,N,M,NQ,QX(10,4),QY(10,4),QZ(10,4),
     & ZT(41,61),ZB(41,61),DZZ,L,CP(5000),CV(5000),SL(5000,4),
     & FEN(41,61),CVN(41,61),SUB(41,61),ZT1(41,61),SDER(41,61),
     & NY,G,DX2,DZ,DM,FF,FB,F3,F4,VST(4),FU(4),FD(4),TC(4),FAC
C.....INITIALIZE TO 0
      DO 5 I=1,N
      DO 5 J=1,M
      CVN(I,J)=(0.,0.)
    5 FEN(I,J)=0.
      NEA=0
C.....ENTER MAIN DO LOOP
      DO 10 K=1,L
C.....CHECK FOR OUT OF BOUNDS
      IF (REAL(CP(K)).LT.0.) GOTO 10
      NEA=NEA+1
C.....FIND NODE
      J=REAL(CP(K))/DX+1.5
      I=N-AIMAG(CP(K))/DX+0.5
C.....ADD TO NODE VELOCITY AND NO. OF ELEMENTS
      CVN(I,J)=CVN(I,J)+CV(K)
      FEN(I,J)=FEN(I,J)+1.
   10 CONTINUE
      IF (NEA.EQ.0) L=0
C.....FIND AVERAGES
      DO 20 I=1,N
      DO 20 J=1,M
   20 IF (FEN(I,J).NE.0.) CVN(I,J)=CVN(I,J)/FEN(I,J)
C.....FIND SECOND DERIVATIVES
      DO 30 I=1,N
```

```
      DO 30 J=1,M
      I1=MAX0(1,I-1)
      J1=MAX0(1,J-1)
      I2=MIN0(N,I+1)
      J2=MIN0(M,J+1)
   30 SDER(I,J)=(ZT(I,J)-(ZT(I1,J)+ZT(I2,J)+ZT(I,J1)+ZT(I,J2))/4.)/DX
      RETURN
      END
C-----------------------------------------------------------------------
      SUBROUTINE READDF3
C.....READS FORMATTED DATA FILE
      IMPLICIT COMPLEX (C)
      CHARACTER*72 IT
      CHARACTER*4 WORD
      DIMENSION LIT (800)
      COMMON /BLK1/LBIT(50,41,61)
      COMMON IT,TO,TR,TD,DT,TB,TE,TID,TF,FLD,SWD,VIS,ROU,DV,CVC,CWV,
     & SDI(4),SDD(4),BD(4),FCO(5),
     & TI,LI,CPI(20),CVI(20),SLI(20,4),TEV,LEV,CPEV(20),
     & DX,N,M,NQ,QX(10,4),QY(10,4),QZ(10,4),
     & ZT(41,61),ZB(41,61),DZZ,L,CP(5000),CV(5000),SL(5000,4),
     & FEN(41,61),CVN(41,61),SUB(41,61),ZT1(41,61),SDER(41,61),
     & NY,G,DX2,DZ,DM,FF,FB,F3,F4,VST(4),FU(4),FD(4),TC(4),FAC
      READ (20,1,ERR=10,END=10)
     & IT,TO,TR,TD,DT,TB,TE,TID,TF,FLD,SWD,VIS,ROU,DV,CVC,CWV,
     & (SDI(K),K=1,4),(SDD(K),K=1,4),(BD(K),K=1,4),
     & (FCO(K),K=1,5),TI,LI
    1 FORMAT (A72///8(11X,F8.0/)//5(11X,F8.0/)2(11X,2F8.0/)
     & ///3(11X,4F8.0/),11X,5F8.0///11X,F8.0/11X,I8/)
      IF (LI.GT.0) READ (20,6,ERR=10,END=10)
     & (CPI(K),CVI(K),(SLI(K,L1),L1=1,4),K=1,LI)
    6 FORMAT (1X,8F8.0)
      READ (20,24,ERR=10,END=10) TEV,LEV
   24 FORMAT (//11X,F8.0/11X,I8/)
      IF (LEV.GT.0) READ (20,4,ERR=10,END=10) (CPEV(LEVI),LEVI=1,LEV)
      READ (20,7,ERR=10,END=10) DX,N,M,NQ
    7 FORMAT (//11X,F8.0/11X,I8/11X,I8///11X,I8/)
      IF (NQ.GT.0) READ (20,4,ERR=10,END=10)
     & ((QX(K,L1),QY(K,L1),QZ(K,L1),L1=1,4),K=1,NQ)
    4 FORMAT (1X,3F8.0)
C.....READ SURFACE GRID AND BASEMENT GRID
      READ (20,8,ERR=10,END=10)
    8 FORMAT (/)
      DO 2 I=1,N
    2 READ (20,3,ERR=10,END=10) (ZT(I,J),J=1,M)
    3 FORMAT (1X,10F7.0)
      READ (20,9,ERR=10,END=10) WORD
    9 FORMAT (/1X,A4)
      IF (WORD.NE.'NOBA') GOTO 38
      DO 33 I=1,N
      DO 33 J=1,M
   33 ZB(I,J)=ZT(I,J)
      READ (20,48,ERR=10,END=10) DZZ
   48 FORMAT (/11X,F8.0)
      GOTO 19
   38 DO 12 I=1,N
   12 READ (20,3,ERR=10,END=10) (ZB(I,J),J=1,M)
C.....READ LITHOLOGY
      READ (20,18,ERR=10,END=10) DZZ
   18 FORMAT (//11X,F8.0)
      DO 15 I=1,N
      DO 15 J=1,M
      READ (20,14,ERR=10,END=10) KL,(LIT(JKL),JKL=1,KL)
   14 FORMAT (1X,I3,1X,12(70I1/))
      DO 13 JKL=1,KL
   13 CALL SEDNOD3 (I,J,JKL,LIT(JKL))
   15 CONTINUE
C.....READ FLUID ELEMENTS
   19 READ (20,5,ERR=10,END=10) L
    5 FORMAT (//11X,I8/)
      IF (L.GT.0) READ (20,17,ERR=10,END=10)
     & (CP(K),CV(K),(SL(K,KS),KS=1,4),K=1,L)
```

```
   17 FORMAT (1X,8F8.0)
C.....FIND PARAMETERS
      DZ=DV/DX/DX
      G=9.81
      DX2=DX*DX
      DM=DV*FLD
      FF=VIS
      FB=ROU*FF
      if (f3.eq.0.) f3=0.2e-10
      F4=.750
      DO 23 K1=1,4
      FU(K1)=1.-FCO(K1)
      VST(K1)=AMIN1(1895.*(SDD(K1)-FLD)**0.8*SDI(K1)**1.4,
     & 4.88*SQRT((SDD(K1)-FLD)*SDI(K1)))
      FD(K1)=AMIN1(1.,VST(K1)*DT/DZ)
   23 TC(K1)=1./SDI(K1)*F3
      FAC=(1.-FCO(5))/DX2/(BD(1)/TC(1)+BD(2)/TC(2)+BD(3)/TC(3)+
     & BD(4)/TC(4))
      RETURN
C.....ERRORS
   10 OPEN (24,FILE='for24.d',FORM='FORMATTED')
      rewind(24)
      WRITE (24,101) TO,TR,L
  101 FORMAT (1X,' TIME=',F15.7,' YEARS OF ',F15.7,' YEARS'/
     & ' L=',I6)
      WRITE (24,11)
   11 FORMAT (' SUBROUTINE READDF3 FOUND ERROR IN DATA FILE')
      CLOSE (24)
      STOP
      END
C-----------------------------------------------------------------------
      SUBROUTINE SEDNOD3 (I,J,K,LIT)
C.....CAUSES EROSION OR DEPOSITION TO OCCUR ON NODE
C     I=ROW OF NODE
C     J=COL OF NODE
C     K=CELL, IF 0, NEGATIVE OR > 800, ERROR IS RETURNED
C     LIT=LITHOLOGY: 1, 2, 3, OR 4. IF 0 or NEGATIVE, LIT IS RETURNED
      IMPLICIT COMPLEX (C)
      CHARACTER*72 IT
      COMMON /BLK1/LBIT(50,41,61)
      COMMON IT,TO,TR,TD,DT,TB,TE,TID,TF,FLD,SWD,VIS,RUU,CVC,DV,
     & SDI(4),SDD(4),BD(4),FCO(5),
     & TI,LI,CPI(20),CVI(20),SLI(20,4),TEV,LEV,CPEV(20),
     & DX,N,M,NQ,QX(10,4),QY(10,4),QZ(10,4),
     & ZT(41,61),ZB(41,61),DZZ,L,CP(5000),CV(5000),SL(5000,4),
     & FEN(41,61),CVN(41,61),SUB(41,61),ZT1(41,61),SDER(41,61),
     & NY,G,DX2,DZ,DM,FF,FB,F3,F4,VST(4),FU(4),FD(4),TC(4),FAC
C.....ERODE
      IF (K.LE.800.AND.K.Gt.0.and.lit.ge.0.and.lit.le.4) GOTO 2
      OPEN (24,FILE='for24.d',FORM='FORMATTED')
      rewind(24)
      WRITE (24,11) TO,TR,L
   11 FORMAT (1X,' TIME=',F15.7,' YEARS OF ',F15.7,' YEARS'/
     & ' L=',I6)
      WRITE (24,12) K,lit
   12 FORMAT (' SED. COLUMN < 1 OR > 800, K=',I4,' LIT=',I1)
      CLOSE (24)
      CALL WRITGU3
      STOP
    2 LW=(K+15)/16
      LB=2*K+14-16*LW
      IF (LIT.GT.0) GOTO 1
      LIT=IBITS(LBIT(LW,I,J),LB,2)+1
      RETURN
C.....DEPOSIT
    1 IF((LIT-1)/2.EQ.1) THEN
      LBIT(LW,I,J)=IBSET(LBIT(LW,I,J),LB+1)
      ELSE
      LBIT(LW,I,J)=IBCLR(LBIT(LW,I,J),LB+1)
      ENDIF
      IF(MOD(LIT,2).EQ.0) THEN
      LBIT(LW,I,J)=IBSET(LBIT(LW,I,J),LB)
```

```
      ELSE
      LBIT(LW,I,J)=IBCLR(LBIT(LW,I,J),LB)
      ENDIF
      RETURN
      END
C-----------------------------------------------------------------------
      SUBROUTINE SET3
C.....SETS TIME EXTRAPOLATION
      IMPLICIT COMPLEX (C)
      CHARACTER*72 IT
      COMMON /BLK1/LBIT(50,41,61)
      COMMON IT,T0,TR,TD,DT,TB,TE,TID,TF,FLD,SWD,VIS,ROU,DV,CVC,CWV,
     & SDI(4),SDD(4),BD(4),FCO(5),
     & TI,LI,CPI(20),CVI(20),SLI(20,4),TEV,LEV,CPEV(20),
     & DX,N,M,NQ,QX(10,4),QY(10,4),QZ(10,4),
     & ZT(41,61),ZB(41,61),DZZ,L,CP(5000),CV(5000),SL(5000,4),
     & FEN(41,61),CVN(41,61),SUB(41,61),ZT1(41,61),SDER(41,61),
     & NY,G,DX2,DZ,DM,FF,FB,F3,F4,VST(4),FU(4),FD(4),TC(4),FAC
      DO 1 I=1,N
      DO 1 J=1,M
    1 ZT1(I,J)=ZT(I,J)
      RETURN
      END
C-----------------------------------------------------------------------
      SUBROUTINE WRITDF3
C.....WRITES FORMATTED DATA FILE
      IMPLICIT COMPLEX (C)
      CHARACTER*72 IT
      DIMENSION LIT (800)
      COMMON /BLK1/LBIT(50,41,61)
      COMMON IT,T0,TR,TD,DT,TB,TE,TID,TF,FLD,SWD,VIS,ROU,DV,CVC,CWV,
     & SDI(4),SDD(4),BD(4),FCO(5),
     & TI,LI,CPI(20),CVI(20),SLI(20,4),TEV,LEV,CPEV(20),
     & DX,N,M,NQ,QX(10,4),QY(10,4),QZ(10,4),
     & ZT(41,61),ZB(41,61),DZZ,L,CP(5000),CV(5000),SL(5000,4),
     & FEN(41,61),CVN(41,61),SUB(41,61),ZT1(41,61),SDER(41,61),
     & NY,G,DX2,DZ,DM,FF,FB,F3,F4,VST(4),FU(4),FD(4),TC(4),FAC
C.....WRITE GENERAL DATA
      WRITE (21,1)
     & IT,T0,TR,TD,DT,TB,TE,TID,TF,FLD,SWD,VIS,ROU,DV,CVC,CWV,
     & (SDI(K),K=1,4),(SDD(K),K=1,4),(BD(K),K=1,4),
     & (FCO(K),K=1,5),TI,LI
    1 FORMAT (A72//' RUN PARAMETERS:'/' START TIM=',E8.2,' Y'/
     & ' END TIM  =',E8.2,' Y'/' TIM/DISPL=',E8.2,' Y'/' TIM INCR1=',
     & E8.2,' S'/' TIM FACT1=',E8.2/' TIM INCR2=',E8.2,' Y'/
     & ' TIM IDLE =',E8.2,' Y'/' TIM FACT2=',E8.2//
     & ' GENERAL PHYSICAL PARAMETERS:'/' FLOW DENS=',F8.0,' KG/M3'/
     & ' SEA DENS =',F8.0,' KG/M3'/' FLOW VISC=',E8.2,' NS/M2'/
     & ' ROUGHNESS=',F8.4/' EL VOLUME=',E8.2,' M3'/' CURR(X,Y)=',2E8.2,
     & 'M/S'/' WAVE(X,Y)=',2E8.2,' M2/S'/
     & ' SEDIMENT PARAMETERS:'/15X,'S1',6X,'S2',6X,'S3',6X,'S4',5X,
     & 'BAS'/' DIAMETER =',4E8.2,'  -----  M'/' DENSITY  =',4F8.0,
     & '  -----  KG/M3'/' BAS DECAY=',4F8.3,'  -----'/' COHESION =',
     & 5E8.2//' SOURCES:'/' INTERVAL =',E8.2,' Y'/' # SOURCES=',I8/
     & '    X(M)     Y(M)   XV(M/S) YV(M/S)  S1(M3)  S2(M3)',
     & '  S3(M3)  S4(M3)')
      IF (LI.GT.0) WRITE (21,6)
     & (CPI(K),CVI(K),(SLI(K,L1),L1=1,4),K=1,LI)
    6 FORMAT (1X,4F8.0,4E8.2)
      WRITE (21,20) TEV,LEV
   20 FORMAT (/' EVAPORATION (UNUSED AT PRESENT)'/
     & ' INTERVAL =',E8.2,' Y'/' # OF ELEM=',I8/'    X-POS    Y-POS')
      IF (LEV.GT.0) WRITE (21,6) (CPEV(LEVI),LEVI=1,LEV)
      WRITE (21,7) DX,N,M,NQ
    7 FORMAT (/' TOPOGRAPHY:'/' GRID SIDE=',F8.1,' M'/' NROWS    =',
     & I8/' NCOLS    =',I8//' TECTONICS (UNUSED AT PRESENT):'
     & /' # QUADS  =',I8/'    X(M)     Y(M)  SUBS(M/Y)')
      IF (NQ.GT.0) WRITE (21,4)
     & ((QX(K,L1),QY(K,L1),QZ(K,L1),L1=1,4),K=1,NQ)
    4 FORMAT (1X,2F8.0,E8.2)
C.....WRITE SURFACE GRID AND BASEMENT GRID
      WRITE (21,8)
```

```
    8 FORMAT (/' GRID NODES ELEVATION (SURFACE) (M)')
      DO 2 I=1,N
    2 WRITE (21,3) (ZT(I,J),J=1,M)
    3 FORMAT (1X,10F7.1)
      WRITE (21,19)
   19 FORMAT (/' GRID NODES ELEVATION (BASEMENT) (M)')
      DO 12 I=1,N
   12 WRITE (21,3) (ZB(I,J),J=1,M)
C.....WRITE LITHOLOGY
      WRITE (21,18) DZZ
   18 FORMAT (/' LITHOLOGIC COLUMNS:'/' LCELL DEP=',F8.3)
      DO 15 I=1,N
      DO 15 J=1,M
      KL=INT((ZT(I,J)-ZB(I,J))/DZZ+0.99999)
      DO 13 JKL=1,KL
      LIT(JKL)=0
   13 CALL SEDNOD3 (I,J,JKL,LIT(JKL))
      WRITE (21,14) KL,(LIT(JKL),JKL=1,KL)
   14 FORMAT (1X,I3,1X,12(70I1/))
   15 CONTINUE
C.....WRITE FLUID ELEMENTS
      DO 22 K=1,L
      DO 22 KS=1,4
   22 SL(K,KS)=SL(K,KS)/TB
      WRITE (21,5) L
    5 FORMAT (/' ELEMENT POSITIONS AND VELOCITIES:'/' # OF ELEM=',I8/
     & '   X(M)     Y(M)    XV(M/S) YV(M/S)   S1(M3)   S2(M3)',
     & '   S3(M3)   S4(M3)')
      IF (L.GT.0) WRITE (21,17)
     & (CP(K),CV(K),(SL(K,KS),KS=1,4),K=1,L)
   17 FORMAT (1X,8F8.0)
      RETURN
      END
C-----------------------------------------------------------------------
      SUBROUTINE WRITGU3
C.....WRITES OUTPUT FILE FOR GRAPHICS
      IMPLICIT COMPLEX (C)
      CHARACTER*72 IT
      DIMENSION WRT (100)
      COMMON /BLK1/LBIT(50,41,61)
      COMMON IT,TO,TR,TD,DT,TB,TE,TID,TF,FLD,SWD,VIS,ROU,DV,CVC,CWV,
     & SDI(4),SDD(4),BD(4),FCO(5),
     & TI,LI,CPI(20),CVI(20),SLI(20,4),TEV,LEV,CPEV(20),
     & DX,N,M,NQ,QX(10,4),QY(10,4),QZ(10,4),
     & ZT(41,61),ZB(41,61),DZZ,L,CP(5000),CV(5000),SL(5000,4),
     & FEN(41,61),CVN(41,61),SUB(41,61),ZT1(41,61),SDER(41,61),
     & NY,G,DX2,DZ,DM,FF,FB,F3,F4,VST(4),FU(4),FD(4),TC(4),FAC
C.....INITIALIZE
      CALL FILTER3 (0)
      EA=0.
C.....WRITE
      IF (TO.NE.0.) GOTO 19
C.....WRITE GENERAL DATA
      WRITE (22)
     & IT,TO,TR,TD,DT,TB,TE,TID,TF,FLD,SWD,VIS,ROU,DV,CVC,CWV,
     & (SDI(K),K=1,4),(SDD(K),K=1,4),(BD(K),K=1,4),
     & (FCO(K),K=1,5),TI,LI
      IF (LI.GT.0) WRITE (22)
     & (CPI(K),CVI(K),(SLI(K,L1),L1=1,4),K=1,LI)
      WRITE (22) TEV,LEV
      IF (LEV.GT.0) WRITE (22) (CPEV(LEVI),LEVI=1,LEV)
      WRITE (22) DX,N,M,NQ
      IF (NQ.GT.0) WRITE (22)
     & ((QX(K,L1),QY(K,L1),QZ(K,L1),L1=1,4),K=1,NQ)
   19 WRITE (22) TO
      DO 8 I=1,N
    8 WRITE (22) (ZT(I,J),J=1,M)
      DO 2 I=1,N
    2 WRITE (22) (ZB(I,J),J=1,M)
C.....WRITE LITHOLOGY
      WRITE (22) DZZ
      DO 9 I=1,N
```

```
      DO 9 J=1,M
      KL=(INT((ZT(I,J)-ZB(I,J))/DZZ+0.99999)+15)/16
    9 WRITE (22) KL,(LBIT(K,I,J),K=1,KL)
C.....WRITE  # OF FLUID ELEMENTS
      DO 10 I=1,N
      DO 29 J=1,M
      EA=EA+FEN(I,J)
   29 WRT(J)=FEN(I,J)
   10 WRITE (22) (WRT(J),J=1,M)
      WRITE (22) EA
      DO 14 I=1,N
      DO 13 J=1,M
   13 WRT(J)=REAL(CVN(I,J))
   14 WRITE (22) (WRT(J),J=1,M)
      DO 18 I=1,N
      DO 17 J=1,M
   17 WRT(J)=AIMAG(CVN(I,J))
   18 WRITE (22) (WRT(J),J=1,M)
      RETURN
      END
C-----------------------------------------------------------------------
```

APPENDIX C

Listing of Program SEDSHO3

Program SEDSHO3 produces the graphic output for the multiple-sediment model, SEDSIM3. A listing of the program and its subroutines is provided. There are subroutines that are hardware-dependent and that are utilized by SEDSHO3, but they are not included in the listing:

OPENPLOT opens a plot and sets plotting scales

PLOT draws a line or moves a pen

OPENPANL opens a panel or color-filled polygon

CLOSPANL closes a panel or color-filled polygon

CHARS draws characters

CLOSPLOT closes a plot

The user should replace the calls to these subroutines by their equivalents on the particular system employed.

```
C     PROGRAM SEDSHO3
      IMPLICIT COMPLEX (C)
      CHARACTER*1 ANS,AN(6),BS
      CHARACTER*8 BLANK,FIL1,FILN
      CHARACTER*72 IT
      COMMON /BLK1/LBIT(50,41,61)
      COMMON IT,TO,TR,TD,DT,TB,TE,TID,TF,FLD,SWD,VIS,ROU,DV,CVC,CWV,
     & SDI(4),SDD(4),BD(4),FCO(5),
```

```
     & TI,LI,CPI(20),CVI(20),SLI(20,4),TEV,LEV,CPEV(20),
     & DX,N,M,NQ,QX(10,4),QY(10,4),QZ(10,4),
     & ZT(41,61),ZB(41,61),DZZ,L,
     & FEN(41,61),CVN(41,61),SUB(41,61),ZT1(41,61),NY
     & /MAP/IPLT,IPEN,IPT,NROWS,NCOLS,
     & ZZ(82,122),XLEN,YLEN,ZSCL,SCLH,SCLV,X0,Y0,Z0,AZ,EL,DS,REF
      DATA BLANK/'        '/
      SAVE /MAP/
C....INITIALIZE
      DO 5 JK=1,5
    5 AN(JK)='Y'
      AN(3)='N'
      AN(6)='N'
      FIL1='for22.d '
      UA=1.
      UD=0.
      REF=0.
      VEX=10.
      NF=1
      AZ=45.
      EL=30.
      NS=1
      XPOS=0.
      YPOS=0.
      STRK=45.
      IFG=0
      FILN=BLANK
      FENMIN=0.00001
      NDC=0
      BS=CHAR(8)
  500 FORMAT (A1)
  501 FORMAT (F10.0)
C.....SET PARAMETERS
    2 WRITE (6,601)
  601 FORMAT (' CHANGE PARAMETERS?(Yes/No/Quit)=',$)
      READ (5,500) ANS
      IF (ANS.EQ.'Q'.OR.ANS.EQ.'q') STOP
      IF (ANS.NE.'Y'.AND.ANS.NE.'y') GOTO 102
      WRITE (6,502) FIL1,(BS,LL=1,8)
  502 FORMAT (' DATA FILE = ',A8,8A1,$)
      READ (5,503) FILN
  503 FORMAT (A8)
      WRITE (6,508) UA,UD,(BS,LL=1,11)
  508 FORMAT (' UNIT CURRENT (M/S), UNIT DEPTH (M) =',2F6.1,11A1,$)
      READ (5,*  ,ERR=21) UA,UD
   21 WRITE (6,505) VEX,REF,(BS,LL=1,11)
  505 FORMAT (' VERT. EXAGGERATION, REF. LEVEL =',F5.1,F7.1,11A1,$)
      READ (5,*  ,ERR=22) VEX,REF
   22 WRITE (6,512) NF,AZ,EL,(BS,LL=1,10)
  512 FORMAT (' PRSPECTIVE #, AZIMUTH, ELEVATION = ',I1,F5.0,F4.0,
     & 10A1,$)
      READ (5,*  ,ERR=210) NF,AZ,EL
      IF (NF.GT.2) GOTO 22
      IF (NF.LT.1) GOTO 22
  210 WRITE (6,510) NS,XPOS,YPOS,STRK,(BS,LL=1,20)
  510 FORMAT (' SECTION #, X-POS, Y-POS, STRIKE = ',I1,2F7.0,F5.0,
     & 20A1,$)
      READ (5,*  ,ERR=102) NS,XPOS,YPOS,STRK
      WRITE (6,514) THCK,(BS,ll=1,6)
  514 FORMAT (' FACIES THICKNESS =',f6.2,6a1,$)
      READ (5,515) THCK
  515 FORMAT (f10.0)
      IF (NS.GT.2) GOTO 210
      IF (NS.LT.1) GOTO 210
C.....OPEN FILE IF NECESSARY
  102 IF ((FILN.EQ.BLANK.OR.FILN.EQ.FIL1).AND.IFG.EQ.1) GOTO 101
      IF (FILN.NE.BLANK.AND.FILN.NE.FIL1) FIL1=FILN
      OPEN (22,FILE=fil1,FORM='UNFORMATTED')
      REWIND (22)
      IFG=1
C.....SET SURFACE DISPLAYS
  101 WRITE (6,602)
```

```
  602 FORMAT (' CHANGE DISPLAYS?(Y/N/Q)=',$)
      READ (5,500) ANS
      IF (ANS.EQ.'Q'.OR.ANS.EQ.'q') STOP
      IF (ANS.NE.'Y'.AND.ANS.NE.'y') GOTO 300
      WRITE (6,603) AN(1),BS
  603 FORMAT ('   1. CONTOUR MAP?(Y/N)=',2A1,$)
      READ (5,500) ANS
      IF (ANS.NE.' ') AN(1)=ANS
      WRITE (6,604) AN(2),BS
  604 FORMAT ('   2. VELOCITY AND DEPTH PLOT?(Y/N)=',2A1,$)
      READ (5,500) ANS
      IF (ANS.NE.' ') AN(2)=ANS
      WRITE (6,605) AN(3),BS
  605 FORMAT ('   3. FISH-NET DIAGRAM?(Y/N)=',2A1,$)
      READ (5,500) ANS
      IF (ANS.NE.' ') AN(3)=ANS
      WRITE (6,606) AN(4),BS
  606 FORMAT ('   4. AGE CROSS SECTION?(Y/N)=',2A1,$)
      READ (5,500) ANS
      IF (ANS.NE.' ') AN(4)=ANS
      WRITE (6,607) AN(5),BS
  607 FORMAT ('   5. SED-TYPE CROSS SECTION?(Y/N)=',2A1,$)
      READ (5,500) ANS
      IF (ANS.NE.' ') AN(5)=ANS
      WRITE (6,610) AN(6),BS
  610 FORMAT ('   6. FACIES MAPS?(Y/N)=',2A1,$)
      READ (5,500) ANS
      IF (ANS.NE.' ') AN(6)=ANS
  300 WRITE (6,608)
  608 FORMAT (' STARTING TIME (YR)=',$)
      READ (5,501,ERR=300) TNDS
      WRITE (6,609)
  609 FORMAT (' ENDING TIME (YR)=',$)
      READ (5,501,ERR=300) TNDE
      IF (TNDE.EQ.0..AND.TNDS.EQ.0.) GOTO 2
C.....PROCESS
  310 REWIND (22)
      NDC=0
  312 CALL READGU3 (IF1,0)
      IF (IF1.EQ.1) GOTO 2
      NY=-1
  350 CALL READGU3 (IF1,1)
      TR=TNDE
      NY=NY+1
      IF (IF1.EQ.1.OR.T0.GT.TNDE*1.0001) GOTO 2
  325 NDC=NDC+1
      IF (T0.LT.TNDS) GO TO 350
  330 IF (AN(1).EQ.'Y'.OR.AN(3).EQ.'Y'.OR.AN(1).EQ.'y'.OR.AN(3).EQ.'y')
     & CALL MAPCYC3 (2,1,0,VEX)
      IPLT=100+NY
      IPEN=3
      IF (AN(1).NE.'Y'.AND.AN(1).NE.'y') GO TO 335
      IF (NY.EQ.0) CALL PLTGRD (1,1)
      CALL MAP2D
  335 IPLT=200+NY
      IF (AN(2).EQ.'Y'.OR.AN(2).EQ.'y')
     & CALL PLTNOD3 (IPLT,5,0,FENMIN,UA,UD)
      IPLT=200+NF*100+NY
      IPEN=6
      IF (AN(3).EQ.'Y'.OR.AN(3).EQ.'y') CALL MAP3D
      IFL=0
      IF (TNDE-T0.LT.TD) IFL=1
      IF (AN(4).EQ.'Y'.OR.AN(4).EQ.'y')
     & CALL CRSSEC3 (NS,VEX,XPOS,YPOS,STRK,REF,0,IFL,IF1)
      IF (AN(5).EQ.'Y'.OR.AN(5).EQ.'y')
     & CALL CRSSEC3 (NS,VEX,XPOS,YPOS,STRK,REF,1,IFL,IF1)
      IF (AN(6).EQ.'Y'.OR.AN(6).EQ.'y') CALL FACIES3 (THCK)
      IF (IF1.EQ.1) GOTO 210
      WRITE (6,199) T0
  199 FORMAT (' TIME =',F15.7,' YEARS')
      GOTO 350
      END
```

```
C-----------------------------------------------------------------------
      SUBROUTINE CRSSEC3 (NS,VEX,XPOS,YPOS,STRK1,REF,LITAGE,IFL,IF1)
C.....DRAWS CROSS SECTION SHOWING AGE OF SEDIMENT
C     NS = SECTION NUMBER (1 OR 2)
C     VEX = VERTICAL EXAGGERATION FOR CROSS SECTION
C     LITAGE = 0 DRAW AGE, = 1 DRAW SEDIMENT TYPE
C     IFL = 1 IF DISPLAY IS LAST IN SEQUENCE
C     SLEN = SECTION LENGTH (METERS, OBJECT SCALE)
C     SLENM = SECTION LENGTH (INCHES, MAP SCALE)
C     XSEC = X POSITION IN SECTION (INCHES, MAP SCALE)
C     YSEC = Y (VERTICAL) POSITION IN SECTION (INCHES, MAP SCALE)
C     YSEC1 = SAME AS YSEC, BUT ONE SURFACE DOWN
C     DZZ = HEIGHT OF ONE LITHOLOGY BLOCK
      IMPLICIT COMPLEX (C)
      CHARACTER*72 IT
      CHARACTER*2 AB(4)
      CHARACTER*6 LEG
      DIMENSION IND(11),XS(2),YS(2),FLIT(4),YSEC(101),YSEC1(101)
C.....FOR STANDARD FORTRAN CHANGE 'EXTENDED BLOCK' TO 'COMMON',
      COMMON /BLK1/LBIT(50,41,61)
      COMMON IT,T0,TR,TD,DT,TB,TE,TID,TF,FLD,SWD,VIS,ROU,DV,CVC,CWV,
     & SDI(4),SDD(4),BD(4),FCO(5),
     & TI,LI,CPI(20),CVI(20),SLI(20,4),TEV,LEV,CPEV(20),
     & DX,N,M,NQ,QX(10,4),QY(10,4),QZ(10,4),
     & ZT(41,61),ZB(41,61),DZZ,L,
     & FEN(41,61),CVN(41,61),SUB(41,61),ZT1(41,61),NY
      DATA IND/-2,-3,-4,-5,-6,-7,-8,-9,-12,-14,-15/ ,PI/3.1415926/,
     & AB/'A ','A''','B','B'''/
      TIM=TR-T0
      IF1=0
      D=2.1*(1-LITAGE)
C.....SKIP IF DISPLAY NOT 0
      IF (NY.NE.0) GOTO 35
C.....INITIALIZE
      E=DX/1000.
      ND=100
      XLEN=6.87/MAX0(N-1,M-1)
      YLEN=XLEN
      ZSCL=XLEN/DX*VEX
      XLE=2.*DX/VEX
      XD=0.25
      YD=0.25
      FLIT(1)=0.
      FLIT(2)=0.
      FLIT(3)=0.
      FLIT(4)=0.
      SHFT=-REF*ZSCL
C.....CONVERT STRIKE TO 1ST HALF AND TO RADIANS AND ROTATE BY PI
      STRK=AMOD(STRK1,180.)
      IF (STRK.LT.0.) STRK=STRK+180.
      IF (STRK.EQ.0.) STRK=0.01
      IF (STRK.EQ.90.) STRK=90.01
      STRK=(90.-STRK)/180.*PI
C.....FIND END POINTS OF SECTION XS(1),YS(1),XS(2),YS(2)
      XMAX=(M-1)*DX-E
      YMAX=(N-1)*DX-E
      TANS=TAN(STRK)
      X1=XPOS-YPOS/TANS
      Y1=YPOS-XPOS*TANS
      X2=XPOS+(YMAX-YPOS)/TANS
      Y2=YPOS+(XMAX-XPOS)*TANS
      I=1
      IF (Y1.LE.0..OR.Y1.GT.YMAX) GOTO 11
      XS(I)=E
      YS(I)=Y1
      I=2
   11 IF (X2.LE.0..OR.X2.GT.XMAX) GOTO 12
      XS(I)=X2
      YS(I)=YMAX
      IF (I.EQ.2) GOTO 44
      I=2
   12 IF (X1.LT.0..OR.X1.GE.XMAX) GOTO 13
```

```
      XS(I)=X1
      YS(I)=E
      IF(I.EQ.2) GOTO 44
      I=2
   13 XS(I)=XMAX
      YS(I)=Y2
      IF (I.EQ.2) GOTO 44
      WRITE (6,17)
   17 FORMAT (' SECTION NOT IN AREA, TRY AGAIN')
      IF1=1
      RETURN
C.....REVERSE POINTS IF NECESSARY
   44 IF (XS(2).GT.XS(1)) GOTO 45
      XT=XS(1)
      YT=YS(1)
      XS(1)=XS(2)
      YS(1)=YS(2)
      XS(2)=XT
      YS(2)=YT
C.....FIND CONSTANTS
   45 SLEN=SQRT((XS(1)-XS(2))**2+(YS(1)-YS(2))**2)
      DSEC=SLEN/ND
      SLENM=SLEN/DX*XLEN
      XSECU=SLENM/ND
C.....DRAW LINE ON MAP
      IPLT=13+(NS*2+LITAGE)
      CALL OPENPLOT (IPLT,1,13.75,13.75,-1.,15.,-5.,11.)
      CALL PLOT (IPLT,XS(1)/DX*XLEN,YS(1)/DX*YLEN,0)
      CALL PLOT (IPLT,XS(2)/DX*XLEN,YS(2)/DX*YLEN,1)
C.....DRAW LABELS AT ENDS OF SECTION LINE
      X=XS(1)/DX*XLEN+0.1*COS(STRK+1.25*PI)
      Y=YS(1)/DX*YLEN+0.1*SIN(STRK+1.25*PI)
      CALL PLOT (IPLT,X,Y,0)
      CALL CHARS (IPLT,STRK*180./PI,0.1,0.,0.,1.,1,2,AB(2*NS-1))
      X=XS(2)/DX*XLEN+0.071*COS(STRK-PI/2.)
      Y=YS(2)/DX*YLEN+0.071*SIN(STRK-PI/2.)
      CALL PLOT (IPLT,X,Y,0)
      CALL CHARS (IPLT,STRK*180./PI,0.1,0.,0.,1.,1,2,AB(2*NS))
      CALL CLOSPLOT (IPLT)
C.....DRAW INITIAL PANEL
   30 IPLT=300+(NS*2+LITAGE)*100
      CALL OPENPLOT (IPLT,1,13.75,13.75,
     & -1./2.,15./2.,-0.1-D,(-0.1-D+16.)/2.+(-0.1-D)/2.)
      CALL PLOT (IPLT,0.,0.,0)
      CALL OPENPANL (1,1)
      CALL PLOT (IPLT,SLENM,0.,1)
      CALL PLOT (IPLT,SLENM,2.,1)
      CALL PLOT (IPLT,0.,2.,1)
      CALL CLOSPANL
C.....DRAW LABELS AT ENDS OF SECTION PANEL
      CALL PLOT (IPLT,-0.1,0.,0)
      CALL CHARS (IPLT,0.,0.1,0.,0.,1.,1,2,AB(2*NS-1))
      CALL PLOT (IPLT,SLENM,0,0)
      CALL CHARS (IPLT,0.,0.1,0.,0.,1.,1,2,AB(2*NS))
C.....ENCODE SCALE
      write (leg,19) XLE
   19 FORMAT (F5.1,'m')
C.....DRAW VERTICAL SCALE
  102 CALL PLOT (IPLT,-0.15,XD+1.-YLEN,0)
      CALL PLOT (IPLT,-0.15,XD+1.+YLEN,1)
      CALL PLOT (IPLT,-.20,XD+0.7,0)
      CALL CHARS (IPLT,90.,0.1,0.,0.,1.,1,6,LEG)
      CALL PLOT (IPLT,-.35,XD+0.4,0)
      CALL CHARS (IPLT,90.,0.1,0.,0.,1.,1,11,'Vert. Scale')
      CALL PLOT (IPLT,SLENM+0.1,0.3,0)
      IF (LITAGE.EQ.0) CALL CHARS (IPLT,0.,0.1,0.,0.,1.,1,3,'AGE')
      IF (LITAGE.EQ.1) CALL CHARS (IPLT,0.,0.1,0.,0.,1.,1,3,'SED')
C.....FIND YSEC AND YSEC1 ARRAYS
   35 DO 215 KD=1,ND+1
      X=XS(1)+(KD-1)*DSEC*COS(STRK)
      Y=YS(1)+(KD-1)*DSEC*SIN(STRK)
C.....FIND CELL (IC,JC) AND PROPORTIONS OF INFLUENCE (FJ,FI)
```

```
      JC=X/DX+1
      IC=N-INT(Y/DX)-1
      FJ=X/DX-INT(X/DX)
      FI=1.-(Y/DX-INT(Y/DX))
      YSEC(KD)=((ZT(IC,JC)*(1-FI)
     &         +ZT(IC+1,JC)*FI)*(1-FJ)
     &        +(ZT(IC,JC+1)*(1-FI)
     &       +ZT(IC+1,JC+1)*FI)*FJ)*ZSCL
      YSEC1(KD)=((ZT1(IC,JC)*(1-FI)
     &          +ZT1(IC+1,JC)*FI)*(1-FJ)
     &         +(ZT1(IC,JC+1)*(1-FI)
     &        +ZT1(IC+1,JC+1)*FI)*FJ)*ZSCL
  215 CONTINUE
C.....OPEN AGE OR SED-TYPE PLOT
      IPLT=300+(NS*2+LITAGE)*100+NY
      IF (NY.GT.0)
     & CALL OPENPLOT (IPLT,1,13.75,13.75,
     & -1./2.,15./2.,-0.1-D,(-0.1-D+16.)/2.+(-0.1-D)/2.)
      IF (LITAGE.EQ.0) GOTO 135
C.....START SED-TYPE PLOT
      DO 15 KD=1,ND
      IF (YSEC(KD).LE.YSEC1(KD).AND.YSEC(KD+1).LE.YSEC1(KD+1)) GOTO 15
      FLIT(1)=0.
      FLIT(2)=0.
      FLIT(3)=0.
      FLIT(4)=0.
      DO 14 ID=-1,0
C.....FIND LITHOLOGY AND COLOR (COLOR INDEX IS GRAPHIC SOFTW. DEPEND.)
      X=XS(1)+(KD+ID)*DSEC*COS(STRK)
      Y=YS(1)+(KD+ID)*DSEC*SIN(STRK)
C.....FIND CELL (IC,JC) AND (FJ,FI) FOR SEGMENT END
      JC=X/DX+1
      IC=N-INT(Y/DX)-1
      FJ=X/DX-INT(X/DX)
      FI=1.-(Y/DX-INT(Y/DX))
      DO 14 II=0,1
      DO 14 JJ=0,1
      ZH=(ZT(IC+II,JC+JJ)-ZB(IC+II,JC+JJ))/DZZ
      ZH1=(ZT1(IC+II,JC+JJ)-SUB(IC+II,JC+JJ)*TD-ZB(IC+II,JC+JJ))/DZZ
      IF (ZH1.GE.ZH.OR.ZH.LE.0.) GOTO 14
      NH=ZH+1
      NH1=ZH1+1
      FACT=(1-II+(2*II-1)*FI)*(1-JJ+(2*JJ-1)*FJ)/(NH-NH1+1)
      DO 24 KH=NH1,NH
      LIT=0
      IF (KH.LE.0) KH=1
      CALL SEDNO13 (IC+II,JC+JJ,KH,LIT)
      FLIT(LIT)=FLIT(LIT)+FACT
   24 CONTINUE
   14 CONTINUE
      FLITOT=FLIT(1)+FLIT(2)+FLIT(3)+FLIT(4)
      IF (FLITOT.LE.0.1) GOTO 15
      IMAX=5-INT(FLIT(4)/FLITOT*4.9999)
      INDCOL=111
      IF (IMAX.EQ.1) GOTO 16
      IND1=MIN0(IMAX,INT(FLIT(1)/AMAX1(FLIT(1)+FLIT(2),FLIT(1)+FLIT(3))
     & *(2*IMAX-1.0001)+1))
      IND2=MIN0(IMAX,INT(FLIT(2)/AMAX1(FLIT(1)+FLIT(2),FLIT(2)+FLIT(3))
     & *(2*IMAX-1.0001)+1))
      IND3=MIN0(IMAX,INT(FLIT(3)/AMAX1(FLIT(1)+FLIT(3),FLIT(2)+FLIT(3))
     & *(2*IMAX-1.0001)+1))
      INDCOL=IND1*100+IND2*10+IND3
C.....DRAW SEDIMENT TYPE PANELS
   16 CALL PLOT (IPLT,XSECU*(KD-1),YSEC(KD)+SHFT,0)
      CALL OPENPANL (INDCOL,0)
      CALL PLOT (IPLT,XSECU*(KD-1),YSEC1(KD)+SHFT,0)
      CALL PLOT (IPLT,XSECU*KD,YSEC1(KD+1)+SHFT,0)
      CALL PLOT (IPLT,XSECU*KD,YSEC(KD+1)+SHFT,0)
      CALL CLOSPANL
   15 CONTINUE
C.....DRAW AGE PANELS OR LITH MASKING PANEL
  135 CALL PLOT (IPLT,0.,2.,0)
```

```
      ICC=MOD(NY,10)+1
      IF (IFL.EQ.1.OR.LITAGE.EQ.1) ICC=11
      CALL OPENPANL (IND(ICC),1)
      DO 115 KD=1,ND+1
  115 CALL PLOT (IPLT,XSECU*(KD-1),YSEC(KD)+SHFT,1)
      CALL PLOT (IPLT,SLENM,2.,0)
      CALL CLOSPANL
      CALL CLOSPLOT (IPLT)
C.....DRAW SEA LEVEL
      IF (SHFT.LE.0..OR.NY.NE.0) RETURN
      IPLT=30+(NS*2+LITAGE)*10
      CALL OPENPLOT (IPLT,1,13.75,13.75,
     & -1./2.,15./2.,-0.1-D,(-0.1-D+16.)/2.+(-0.1-D)/2.)
      CALL PLOT (IPLT,0.,SHFT,0)
      CALL PLOT (IPLT,SLENM,SHFT,5)
      CALL CLOSPLOT (IPLT)
      RETURN
      END
C----------------------------------------------------------------------
      SUBROUTINE DRBOTT
      COMMON /MAP/IPLT,IPEN,IPT        ,NROWS,NCOLS,Z(82,122),XLEN,
     & YLEN,ZSCL,SCLH,SCLV,X0,Y0,Z0,AZ,EL,DS,REF
     & /PERS/SINAZ,COSAZ,SINEL,COSEL
      SAVE /MAP/,/PERS/
      DO 1 K=1,4
      JINC=(K-1)/2
      IINC=MOD(K,4)/2
      IP=(K+2)/4*IPEN
      X1=JINC*XLEN*NCOLS
      Y1=IINC*YLEN*NROWS
      CALL PERSP (X1,Y1,REF,X2,Y2,Z2,IFLAG)
      CALL PLOT (IPLT,X2,Y2,IP)
      IF (K.EQ.1) CALL OPENPANL (-15,1)
    1 IF (K.EQ.4) CALL CLOSPANL
      RETURN
      END
C----------------------------------------------------------------------
      SUBROUTINE DRCELL(I,J)
      COMMON /MAP/IPLT,IPEN,IPT        ,NROWS,NCOLS,Z(82,122),XLEN,
     & YLEN,ZSCL,SCLH,SCLV,X0,Y0,Z0,AZ,EL,DS,REF
     & /PERS/SINAZ,COSAZ,SINEL,COSEL
      SAVE /MAP/,/PERS/
      DO 1 K=1,4
      JINC=(K-1)/2
      IINC=1-MOD(K,4)/2
      IP=(K+2)/4*IPEN
      X1=(J-1+JINC)*XLEN
      Y1=(NROWS-I-IINC)*YLEN
      Z1=Z(I+IINC,J+JINC)
      CALL PERSP (X1,Y1,Z1,X2,Y2,Z2,IFLAG)
      CALL PLOT (IPLT,X2,Y2,IP)
      IF (K.EQ.1) CALL OPENPANL (-15,1)
    1 IF (K.EQ.4) CALL CLOSPANL
      RETURN
      END
C----------------------------------------------------------------------
      SUBROUTINE DRSCAL
C.DRAWS SCALE
      COMMON /MAP/IPLT,IPEN,IPT        ,NROWS,NCOLS,Z(82,122),XLEN,
     & YLEN,ZSCL,SCLH,SCLV,X0,Y0,Z0,AZ,EL,DS,REF
     & /PERS/SINAZ,COSAZ,SINEL,COSEL
      SAVE /MAP/,/PERS/
      CHARACTER*6 STRNG(3)
      LSCLH=SCLH*4
      LSCLV=SCLV*4
      WRITE (STRNG,10) LSCLH,LSCLH,LSCLV
   10 FORMAT (2(I5,'m'/)I4,'m ')
      CALL PERSP ((NCOLS+7)*XLEN,((NROWS-1)/2-2)*YLEN,0.,
     & XP1,YP1,ZP1,IFLAG)
      CALL PERSP ((NCOLS+7)*XLEN,((NROWS-1)/2+2)*YLEN,0.,
     & XP2,YP2,ZP2,IFLAG)
      CALL PERSP ((NCOLS+3)*XLEN,((NROWS-1)/2-2)*YLEN,0.,
```

```
     & XP3,YP3,ZP3,IFLAG)
      CALL PERSP ((NCOLS+7)*XLEN,((NROWS-1)/2-2)*YLEN,4.*XLEN/ZSCL,
     & XP4,YP4,ZP4,IFLAG)
      CALL PLOT (IPLT,XP1,YP1,0)
      CALL PLOT (IPLT,XP2,YP2,1)
      CALL CHARS (IPLT,0.,0.1,0.,0.,1.,1,6,STRNG(1))
      CALL PLOT (IPLT,XP1,YP1,0)
      CALL PLOT (IPLT,XP3,YP3,1)
      CALL CHARS (IPLT,0.,0.1,0.,0.,1.,1,6,STRNG(2))
      CALL PLOT (IPLT,XP1,YP1,0)
      CALL PLOT (IPLT,XP4,YP4,1)
      CALL CHARS (IPLT,0.,0.1,0.,0.,1.,1,6,STRNG(3))
      RETURN
      END
C-----------------------------------------------------------------------
      SUBROUTINE DRSIDE (IXY,N)
C.DRAWS SIDE OR SLICE OF BLOCK DIAGRAM
C  IXY=0 TO DRAW X-AXIS SIDE, 1 TO DRAW Y-AXIS SIDE
C  N=ROW OR COLUMN NUMBER
      COMMON /MAP/IPLT,IPEN,IPT        ,NROWS,NCOLS,Z(82,122),XLEN,
     & YLEN,ZSCL,SCLH,SCLV,X0,Y0,Z0,AZ,EL,DS,REF
     & /PERS/SINAZ,COSAZ,SINEL,COSEL
      SAVE /MAP/,/PERS/
      IF (IXY.EQ.1) GOTO 2
      Y1=(NROWS-N)*YLEN
      CALL PERSP (0.,Y1,REF,X2,Y2,Z2,IFLAG)
      CALL PLOT (IPLT,X2,Y2,0)
      CALL OPENPANL(-15,1)
      DO 1 J=1,NCOLS
      X1=(J-1)*XLEN
      Z1=Z(N,J)
      CALL PERSP (X1,Y1,Z1,X2,Y2,Z2,IFLAG)
    1 CALL PLOT (IPLT,X2,Y2,IPEN)
      CALL PERSP (X1,Y1,REF,X2,Y2,Z2,IFLAG)
      CALL PLOT (IPLT,X2,Y2,IPEN)
      CALL CLOSPANL
      RETURN
    2 X1=(N-1)*YLEN
      Y1=(NROWS-1)*YLEN
      CALL PERSP (X1,Y1,REF,X2,Y2,Z2,IFLAG)
      CALL PLOT (IPLT,X2,Y2,0)
      CALL OPENPANL (-15,1)
      DO 3 I=1,NROWS
      Y1=(NROWS-I)*XLEN
      Z1=Z(I,N)
      CALL PERSP (X1,Y1,Z1,X2,Y2,Z2,IFLAG)
    3 CALL PLOT (IPLT,X2,Y2,IPEN)
      CALL PERSP (X1,Y1,REF,X2,Y2,Z2,IFLAG)
      CALL PLOT (IPLT,X2,Y2,IPEN)
      CALL CLOSPANL
      RETURN
      END
C-----------------------------------------------------------------------
      SUBROUTINE FACIES(THCK)
C.....DRAWS FACIES MAP
C     THCK= THICKNESS TO BE DRAWN, IF 0 LAST CYCLE'S
C            DEPOSIT IS DRAWN
C     DZZ = HEIGHT OF ONE LITHOLOGY BLOCK
      IMPLICIT COMPLEX (C)
      CHARACTER*72 IT
      DIMENSION FLIT(4)
      COMMON /BLK1/LBIT(50,41,61)
      COMMON IT,T0,TR,TD,DT,TB,TE,TID,TF,FLD,SWD,VIS,ROU,DV,CVC,CWV,
     & SDI(4),SDD(4),BD(4),FCO(5),
     & TI,LI,CPI(20),CVI(20),SLI(20,4),TEV,LEV,CPEV(20),
     & DX,N,M,NQ,QX(10,4),QY(10,4),QZ(10,4),
     & ZT(41,61),ZB(41,61),DZZ,L,
     & FEN(41,61),CVN(41,61),SUB(41,61),ZT1(41,61),NY
      COMMON /PLT/XMN,YMN,XMX,YMX,XSCL,YSCL,LPEN,MS,IXT,IYT
      MS=1
      TIM=TR-T0
      IF1=0
```

```
C.....SKIP IF DISPLAY NOT 0
C.....INITIALIZE
C     NC=CELL MULTIPLICITY
      NC=6
      DNC=1./NC/2.
      XYLEN=6.87/MAX0(N-1,M-1)
      IPLT=0
      CALL OPENPLOT (IPLT,1,13.75,13.75,-1.,15.,-5.,11.)
C.....ENTER MAIN LOOPS
   35 DO 215 IC=1,N-1
      DO 214 JC=1,M-1
      DO 213 ID=1,NC
      DO 212 JD=1,NC
      FJ=(JD*2.-1)*DNC
      FI=(ID*2.-1.)*DNC
      Y=(N-IC)-FI
      X=(JC-1)+FJ
C.....FIND CELL (IC,JC) AND PROPORTIONS OF INFLUENCE (FJ,FI)
      YSEC    =((ZT(IC,JC)*(1-FI)
     &         +ZT(IC+1,JC)*FI)*(1-FJ)
     &       +(ZT(IC,JC+1)*(1-FI)
     &     +ZT(IC+1,JC+1)*FI)*FJ)
      YSEC1    =((ZT1(IC,JC)*(1-FI)
     &          +ZT1(IC+1,JC)*FI)*(1-FJ)
     &        +(ZT1(IC,JC+1)*(1-FI)
     &     +ZT1(IC+1,JC+1)*FI)*FJ)
C.....START SED-TYPE CELL
      IF (YSEC.LE.YSEC1.AND.THCK.LE.0.) GOTO 212
      FLIT(1)=0.
      FLIT(2)=0.
      FLIT(3)=0.
      FLIT(4)=0.
C.....FIND LITHOLOGY AND COLOR (COLOR INDEX IS GRAPHIC SOFTW. DEPEND.)
      DO 15 II=0,1
      DO 14 JJ=0,1
      ZH=(ZT(IC+II,JC+JJ)-ZB(IC+II,JC+JJ))/DZZ
      if (thck.le.0)
     &ZH1=(ZT1(IC+II,JC+JJ)-SUB(IC+II,JC+JJ)*TD-ZB(IC+II,JC+JJ))/DZZ
      IF (THCK.GT.0.) ZH1=ZH-THCK/DZZ
      IF (ZH1.LT.0.) ZH1=0.
      IF (ZH1.GE.ZH.OR.ZH.LE.0.) GOTO 14
      NH=ZH+1
      NH1=ZH1+1
      FACT=(1-II+(2*II-1)*FI)*(1-JJ+(2*JJ-1)*FJ)/(NH-NH1+1)
      IF (NH.LE.0) NH=1
      DO 24 KH=NH1,NH
      lit=0
      CALL SEDNO13 (IC+II,JC+JJ,KH,LIT)
      FLIT(LIT)=FLIT(LIT)+FACT
   24 CONTINUE
   14 CONTINUE
   15 CONTINUE
      FLITOT=FLIT(1)+FLIT(2)+FLIT(3)+FLIT(4)
      IF (FLITOT.LE.0.1) GOTO 212
      IMAX=5-INT(FLIT(4)/FLITOT*4.9999)
      INDCOL=111
      IF (IMAX.EQ.1) GOTO 16
      IND1=MIN0(IMAX,INT(FLIT(1)/AMAX1(FLIT(1)+FLIT(2),FLIT(1)+FLIT(3))
     & *(2*IMAX-1.0001)+1))
      IND2=MIN0(IMAX,INT(FLIT(2)/AMAX1(FLIT(1)+FLIT(2),FLIT(2)+FLIT(3))
     & *(2*IMAX-1.0001)+1))
      IND3=MIN0(IMAX,INT(FLIT(3)/AMAX1(FLIT(1)+FLIT(3),FLIT(2)+FLIT(3))
     & *(2*IMAX-1.0001)+1))
      INDCOL=IND1*100+IND2*10+IND3
C.....DRAW SEDIMENT TYPE CELLS
   16 CALL PLOT (IPLT,(X-DNC)*XYLEN,(Y+DNC)*XYLEN,0)
      CALL OPENPANL (INDCOL,0)
      CALL PLOT (IPLT,(X+DNC)*XYLEN,(Y+DNC)*XYLEN,0)
      CALL PLOT (IPLT,(X+DNC)*XYLEN,(Y-DNC)*XYLEN,0)
      CALL PLOT (IPLT,(X-DNC)*XYLEN,(Y-DNC)*XYLEN,0)
      CALL CLOSPANL
  212 CONTINUE
```

```
  213 CONTINUE
  214 CONTINUE
  215 CONTINUE
      CALL CLOSPLOT (IPLT)
      RETURN
      END
C-----------------------------------------------------------------------
      SUBROUTINE MAP2D
C.....SUBROUTINE TO PLOT A CONTOUR MAP FROM RECTANGULAR GRID.
C     Z = ARRAY CONTAINING Z VALUES
C     TG = ARRAY INDICATING WHETHER SEGMENT IS TAGGED
      CHARACTER*10 LABEL
      CHARACTER*6 STRNG
      LOGICAL TG
      COMMON /MAP/IPLT,IPEN,IPT        ,NROWS,NCOLS,Z(82,122),XLEN,
     & YLEN,ZSCL,SCLH,SCLV,X0,Y0,Z0,AZ,EL,DS,REF,/tag/tg(82,122)
      SAVE /MAP/,/TAG/
C.....INITIALIZE
      DO 6 I=1,NROWS
      DO 6 J=1,NCOLS
    6 TG(i,j)=.FALSE.
C.....OPEN PLOT
      XMAX=(NCOLS-1)*XLEN
      YMAX=(NROWS-1)*YLEN
      CALL OPENPLOT (IPLT,1  ,13.75,13.75,-1.,15.,-5.,11.)
C.....DRAW BOX
      CALL PLOT (IPLT,0.,0.,0)
      CALL OPENPANL (-15,1)
      CALL PLOT (IPLT,XMAX,0.,IPEN)
      CALL PLOT (IPLT,XMAX,YMAX,IPEN)
      CALL PLOT (IPLT,0.,YMAX,IPEN)
      CALL PLOT (IPLT,0.,0.,IPEN)
      CALL CLOSPANL
C.....FIND Z EXTREMA
      ZMAX=Z(1,1)
      ZMIN=ZMAX
      DO 1 I=1,NROWS
      DO 1 J=1,NCOLS
      ZMAX=AMAX1(ZMAX,Z(I,J))
    1 ZMIN=AMIN1(ZMIN,Z(I,J))
C.....FIND CONTOURING INTERVAL
      ZRANG=ZMAX-ZMIN
      IF (ZRANG.EQ.0.) GOTO 110
      CINT=ZRANG/20.
      CMAG=ALOG10(CINT)
      MAG=CMAG
      IF (CMAG.LT.0.) MAG=MAG-1
      CMANT=CMAG-MAG
      IVAL=10.**CMANT+0.5
      IF (IVAL.EQ.3) IVAL=2
      IF (IVAL.GE.4.AND.IVAL.LE.6) IVAL=5
      IF (IVAL.LT.7) GOTO 2
      IVAL=1
      MAG=MAG+1
    2 CINT=10.**MAG*IVAL
      CMIN=INT(ZMIN/CINT)*CINT
      IF (ZMIN.GT.0.) CMIN=CMIN+CINT
      CMAX=INT(ZMAX/CINT)*CINT
      IF (ZMAX.LT.0.) CMAX=CMAX-CINT
      NCON=(CMAX-CMIN)/CINT+1.1
C.....ENTER MAIN DO LOOP, ONCE PER CONTOUR
      DO 100 IC=1,NCON
      CONT=CMIN+CINT*(IC-1)
      write (LABEL,5) CONT
    5 FORMAT (G10.4)
C.....SPECIFY PEN FOR THICK CONTOURS
      IF(MOD(INT(CONT/CINT),5).EQ.0) IPEN=IPEN
C.....ENTER SECONDARY DO LOOPS, ONCE PER HORIZ. SEGMENT, tg SEGMENTS
      DO 30 I=2,NROWS-1
      DO 30 J=1,NCOLS-1
   30 tg(I-1,J)=Z(I,J).LE.CONT.AND.Z(I,J+1).GT.CONT
C.....PLOT CONTOURS
```

```
C.....ENTER 4 INDEPENDENT DO LOOPS (ONCE PER EACH EDGE)
C.....TOP EDGE
      DO 40 J=1,NCOLS-1
      I1=1
      I2=1
      J1=J
      J2=J1+1
      IF (Z(I1,J1).LE.CONT.AND.Z(I2,J2).GT.CONT)
     +CALL PLTCON (CONT,I1,J1,I2,J2,LABEL)
   40 CONTINUE
C.....RIGHT EDGE
      DO 50 I=1,NROWS-1
      J1=NCOLS
      J2=NCOLS
      I1=I
      I2=I1+1
   50 IF (Z(I1,J1).LE.CONT.AND.Z(I2,J2).GT.CONT)
     +CALL PLTCON (CONT,I1,J1,I2,J2,LABEL)
C.....BOTTOM EDGE
      DO 60 J=1,NCOLS-1
      I1=NROWS
      I2=NROWS
      J2=NCOLS-J
      J1=J2+1
   60 IF (Z(I1,J1).LE.CONT.AND.Z(I2,J2).GT.CONT)
     +CALL PLTCON (CONT,I1,J1,I2,J2,LABEL)
C.....LEFT EDGE
      DO 70 I=1,NROWS-1
      J1=1
      J2=1
      I2=NROWS-I
      I1=I2+1
   70 IF (Z(I1,J1).LE.CONT.AND.Z(I2,J2).GT.CONT)
     +CALL PLTCON (CONT,I1,J1,I2,J2,LABEL)
C.....ENTER NESTED DO LOOPS FOR CLOSED CONTOURS
      DO 80 I=2,NROWS-1
      DO 80 J=1,NCOLS-1
      I1=I
      I2=I1
      J1=J
      J2=J1+1
   80 IF (tg(I1-1,J1))
     +CALL PLTCON (CONT,I1,J1,I2,J2,LABEL)
  100 CONTINUE
C.....DRAW SCALE
  110 CALL PLOT (IPLT,0.5,-0.15,0)
      CALL PLOT (IPLT,XLEN*4+0.5,-0.15,1)
      LEG=4*SCLH
      write (STRNG,13) LEG
   13 FORMAT (I5,'m')
      CALL PLOT (IPLT,0.4,-0.30,0)
      CALL CHARS (IPLT,0.,0.1,0.,0.,1.,1,6,STRNG)
      CALL PLOT (IPLT,0.1,-0.45,0)
      CALL CHARS (IPLT,0.,0.1,0.,0.,1.,1,12,'Horiz. Scale')
      CALL PLOT (IPLT,1.8,-0.45,0)
      CALL CHARS (IPLT,0.,0.1,0.,0.,1.,1,18,'Contours in meters')
C.....CLOSE PLOT
      CALL CLOSPLOT (IPLT)
      RETURN
      END
C-----------------------------------------------------------------------
      SUBROUTINE MAP3D
C.....MAKES BLOCK DIAGRAM FROM SQUARE OR RECTANGULAR GRID
      COMMON /MAP/IPLT,IPEN,IPT        ,NROWS,NCOLS,Z(82,122),XLEN,
     & YLEN,ZSCL,SCLH,SCLV,X0,Y0,Z0,AZ,EL,DS,REF
     & /PERS/SINAZ,COSAZ,SINEL,COSEL
      SAVE /MAP/,/PERS/
C.....NROWS,NCOLS= NUMBER OF ROWS AND COLUMNS OF NODES
C     Z= ARRAY CONTAINING Z VALUES
C     X0,Y0,Z0= COORDINATES OF CENTER OF VIEW (INCHES)
C     AZ= AZIMUTH OF OBSERVER (CLOCKWISE FROM Y AXIS (NORTH))
C     EL= ELEVATION OF OBSERVER (UP FROM HORIZONTAL)
```

```
C      DS= DISTANCE FROM OBSERVER TO CENTER OF VIEW (INCHES)
C      XLEN,YLEN= LENGTH OF SIDES OF EACH CELL (INCHES)
C      ZSCL=VERTICAL SCALE IN INCHES PER UNIT
C.....INITIALIZE
       SINAZ=SIN(AZ*3.1416/180.)
       COSAZ=COS(AZ*3.1416/180.)
       SINEL=SIN(EL*3.1416/180.)
       COSEL=COS(EL*3.1416/180.)
C.....FIND POSITION OF OBSERVER
       XOBS=X0+DS*SINAZ*COSEL
       YOBS=Y0+DS*COSAZ*COSEL
       ZOBS=Z0+DS*SINEL
       IOBS=NROWS-INT(YOBS/YLEN+1001.)+1000
       JOBS=INT(XOBS/XLEN+1001.)-1000
C..OPEN PLOT
       CALL OPENPLOT (IPLT,1,13.75,13.75,-16.,5.,-10.5,10.5)
C..ENTER 4 DO LOOPS TO DRAW CORNER AREAS
       DO 10 I=1,MIN0(IOBS-1,NROWS-1)
       DO 10 J=NCOLS-1,MAX0(JOBS+1,1),-1
   10 CALL DRCELL(I,J)
       DO 11 J=NCOLS-1,MAX0(JOBS+1,1),-1
       DO 11 I=NROWS-1,MAX0(IOBS+1,1),-1
   11 CALL DRCELL(I,J)
       DO 12 I=NROWS-1,MAX0(IOBS+1,1),-1
       DO 12 J=1,MIN0(JOBS-1,NCOLS-1)
   12 CALL DRCELL(I,J)
       DO 13 J=1,MIN0(JOBS-1,NCOLS-1)
       DO 13 I=1,MIN0(IOBS-1,NROWS-1)
   13 CALL DRCELL(I,J)
C...DRAW ROW AND COLUMN OF OBSERVER
       IF (JOBS.LT.1.OR.JOBS.GT.NCOLS-1) GOTO 16
       DO 14 I=1,MIN0(IOBS-1,NROWS-1)
   14 CALL DRCELL(I,JOBS)
       DO 15 I=NROWS-1,MAX0(IOBS+1,1),-1
   15 CALL DRCELL(I,JOBS)
   16 IF (IOBS.LT.1.OR.IOBS.GT.NROWS-1) GOTO 19
       DO 17 J=NCOLS-1,MAX0(JOBS+1,1),-1
   17 CALL DRCELL(IOBS,J)
       DO 18 J=1,MIN0(JOBS-1,NCOLS-1)
   18 CALL DRCELL(IOBS,J)
C...DRAW CENTER CELL
   19 IF (JOBS.GE.1.AND.JOBS.LE.NCOLS.AND.IOBS.GE.1.AND.IOBS.LE.NCOLS)
     # CALL DRCELL(IOBS,JOBS)
C...DRAW SIDES
       IF (JOBS.LT.1) CALL DRSIDE (1,1)
       IF (JOBS.GT.NCOLS) CALL DRSIDE (1,NCOLS)
       IF (IOBS.LT.1) CALL DRSIDE (0,1)
       IF (IOBS.GT.NROWS) CALL DRSIDE(0,NROWS)
C...DRAW BOTTOM
       IF (ZOBS.LT.0) CALL DRBOTT
C.....DRAW SCALE
       CALL DRSCAL
C..CLOSE PLOT
       CALL CLOSPLOT (IPLT)
       RETURN
       END
C-----------------------------------------------------------------------
       SUBROUTINE MAPCYC3 (IF,K1,K2,VEX)
C.....BUILDS GRID TO BE MAPPED
C      K1=1 ADD TOPOGRAPHY TO MAPPED QUANTITY  K1=0 DON'T ADD
C      K2=1 ADD WATER DEPTH TO MAPPED QUANTITY, K2=0 DON'T ADD
C      VEX = VERTICAL EXAGGERATION FOR PERSPECTIVE DIAGRAM
C      IF = CELL MULTIPLICITY
       IMPLICIT COMPLEX (C)
       CHARACTER*72 IT
C.....FOR STANDARD FORTRAN CHANGE 'EXTENDED BLOCK' TO 'COMMON',
       COMMON IT,T0,TR,TD,DT,TB,TE,TID,TF,FLD,SWD,VIS,ROU,DV,CVC,CWV,
     & SDI(4),SDD(4),BD(4),FCO(5),
     & TI,LI,CPI(20),CVI(20),SLI(20,4),TEV,LEV,CPEV(20),
     & DX,N,M,NQ,QX(10,4),QY(10,4),QZ(10,4),
     & ZT(41,61),ZB(41,61),DZZ,L,
     & FEN(41,61),CVN(41,61),SUB(41,61),ZT1(41,61),NY,
```

```
     & /MAP/IPLT,IPEN,ipt,NROWS,NCOLS,
     & ZZ(82,122),XLEN,YLEN,ZSCL,SCLH,SCLV,X0,Y0,Z0,AZ,EL,DS,REF
C.....SET CONSTANTS
      SCLH=DX/IF
      SCLV=SCLH/VEX
      NCOLS=(M-1)*IF+1
      NROWS=(N-1)*IF+1
C.....JUMP TO APPROPRIATE SECTION OF PROGRAM
      IF (IF.EQ.0) RETURN
      IF (IF/2) 20,10,10
C.....REDUCE GRID SIDE
   10 DO 1 I3=1,NROWS-1
      DO 1 J3=1,NCOLS-1
      I1=MOD(I3-1,IF)
      J1=MOD(J3-1,IF)
      I2=(I3-1)/IF+1
      J2=(J3-1)/IF+1
      I4=(I3-1+IF/2)/IF+1
      J4=(J3-1+IF/2)/IF+1
    1 ZZ(I3,J3)=(ZT(I2,J2)*(IF-I1)*(IF-J1)+ZT(I2+1,J2)*I1*(IF-J1)+
     #ZT(I2,J2+1)*(IF-I1)*J1+ZT(I2+1,J2+1)*I1*J1)/(IF*IF)
C.....DO LAST COLUMN EXCEPT LAST ELEMENT
      DO 2 I3=1,NROWS-1
      I1=MOD(I3-1,IF)
      I2=(I3-1)/IF+1
      I4=(I3-1+IF/2)/IF+1
    2 ZZ(I3,NCOLS)=(ZT(I2,M)*(IF-I1)+ZT(I2+1,M)*I1)/IF
C.....DO LAST ROW EXCEPT LAST ELEMENT
      DO 3 J3=1,NCOLS-1
      J1=MOD(J3-1,IF)
      J2=(J3-1)/IF+1
      J4=(J3-1+IF/2)/IF+1
    3 ZZ(NROWS,J3)=(ZT(N,J2)*(IF-J1)+ZT(N,J2+1)*J1)/IF
C.....DO LAST ELEMENT
      ZZ(NROWS,NCOLS)=ZT(N,M)
      GOTO 30
C.....INCREASE GRID SIDE
   20 DO 5 I1=1,(N-1)/(-IF)+1
      DO 5 J1=1,(M-1)/(-IF)+1
    5 ZZ(I1,J1)=ZT((I1-1)*(-IF)+1,(J1-1)*(-IF)+1)
C.....MAP
   30 XLEN=6.87/MAX0(NROWS-1,NCOLS-1)
      YLEN=XLEN
      ZSCL=XLEN/DX*VEX*IF
      X0=(NCOLS-1)*XLEN/2.
      Y0=(NROWS-1)*YLEN/2.
      Z0=0.
      DS=(X0+Y0)*2.
C.....AZ MUST DIFFER BY 4 DEGREES BETWEEN VIEW OF EACH EYE
      RETURN
      END
C---------------------------------------------------------------------
      SUBROUTINE PERSP (XI,YI,ZI,XP,YP,ZP,IFLAG)
C.....TRANSFORMS (X,Y,Z) IN SPACE TO (XP,YP) ON A PLANE, ZP IS THE
C       DISTANCE FROM OBSERVER'S PLANE.
C      (X0,Y0,Z0)=COORDINATES OF CENTER OF VIEW (COMMONLY CENTER OF
C                 OBJECT BEING VIEWED)
C      AZ=AZIMUTH OF OBSERVER (CLOCKWISE FROM Y AXIS (NORTH))
C      EL=ELEVATION OF VIEWER (UP FROM HORIZONTAL)
C      DS=DISTANCE OF VIEWER FROM CENTER OF VIEW
C      IFLAG=1 IF ERROR OCCURS (I.E. IF POINT BEHIND VIEWER)
      COMMON /MAP/IPLT,IPEN,IPT        ,NROWS,NCOLS,Z(82,122),XLEN,
     & YLEN,ZSCL,SCLH,SCLV,X0,Y0,Z0,AZ,EL,DS,REF
     & /PERS/SINAZ,COSAZ,SINEL,COSEL
      SAVE /MAP/,/PERS/
      IFLAG=0
C.....TRANSLATE
      X1=XI-X0
      Y1=YI-Y0
      Z1=ZI*ZSCL-Z0
C.....ROTATE AZIMUTH
      X2=-COSAZ*X1+SINAZ*Y1
```

```
      Y2=-SINAZ*X1-COSAZ*Y1
      Z2=Z1
C.....ROTATE ELEVATION
      X3=X2
      Y3=SINEL*Y2+COSEL*Z2
      Z3=-COSEL*Y2+SINEL*Z2
C.....TRANSFORM FOR PERSPECTIVE
      ZP=DS-Z3
      IF (ZP.GT.0.) GOTO 1
      IFLAG=1
      RETURN
    1 C=DS/ZP
      XP=X3*C
      YP=Y3*C
      RETURN
      END
C-----------------------------------------------------------------------
      SUBROUTINE PLTARR (IPLT,X1,Y1,X2,Y2,Z,IPEN)
C.....PLOTS ARROW BETWEEN TWO POINTS
      DATA PI/3.1415926/
      ARG(Z1,Z2)=ATAN(Z2/Z1)+PI*(0.5-SIGN(0.5,Z1))
C.....PLOT ARROW SHAFT
      CALL PLOT (IPLT,X1,Y1,0)
      CALL PLOT (IPLT,X2,Y2,IPEN)
C.....PLOT ARROW HEAD
      IF (X1.EQ.X2.AND.Y2.GE.Y1) ANG=-PI/2.
      IF (X1.EQ.X2.AND.Y2.LT.Y1) ANG=PI/2.
      IF (X1.NE.X2) ANG=ARG(X1-X2,Y1-Y2)
      XP=X2+0.05*COS(ANG+0.4)
      YP=Y2+0.05*SIN(ANG+0.4)
      CALL PLOT(IPLT,XP,YP,0)
      CALL PLOT(IPLT,X2,Y2,IPEN)
      XP=X2+0.05*COS(ANG-0.4)
      YP=Y2+0.05*SIN(ANG-0.4)
      CALL PLOT (IPLT,XP,YP,IPEN)
C.....PLOT DEPTH SEGMENT ACROSS ARROW
      XZ1=(X1+X2-SIN(ANG)*Z)/2
      YZ1=(Y1+Y2+COS(ANG)*Z)/2
      XZ2=XZ1+SIN(ANG)*Z
      YZ2=YZ1-COS(ANG)*Z
      CALL PLOT(IPLT,XZ1,YZ1,0)
      CALL PLOT(IPLT,XZ2,YZ2,IPEN-1)
      RETURN
      END
C-----------------------------------------------------------------------
      SUBROUTINE PLTCON (CONT,I1,J1,I2,J2,LABEL)
C.....PLOTS A SINGLE CONTOUR LINE STARTING AT SEGMENT (I1,J1) (I2,J2)
C.....PLOTTED CONTOUR STARTS TO THE RIGHT OF SEGMENT
C     IDL=1 WHILE LABEL IS BEING DRAWN, 0 ELSE
C     IL=1 WHEN A LABEL HAS ALREADY BEEN DRAWN, 0 ELSE
C     CLV= CONTOUR LENGTH OF LAST VECTOR
C     CCLL= CUM. CONT. LENGTH SINCE LAST LABEL
C     CCLT= CONT LENGTH SINCE START OF CONTOUR (USED FOR TICKS)
C     CLL= CONT. LENGTH OF LABEL
C     CLBL= CONT. LENGTH BETWEEN LABELS
C     CLBT= CONT. LENGTH BETWEEN TICKS
C     XTICK1(I),YTICK1(I),XTICK2(I),YTICK2(I)= COORDS. OF TICKS
C     XTEMP(I),YTEMP(I)= COORDS. OF TEMPORARY VECTORS
      DIMENSION XTICK1(16),YTICK1(16),XTICK2(16),YTICK2(16),XT(16),
     & YT(16)
      CHARACTER*5 LABEL
      LOGICAL TG
      COMMON /MAP/IPLT,IPEN,IPT        ,NROWS,NCOLS,Z(82,122),XLEN,
     & YLEN,ZSCL,SCLH,SCLV,X0,Y0,Z0,AZ,EL,DS,REF,/tag/tg(82,122)
      SAVE /MAP/,/TAG/
C.....SET PARAMETERS AND INITIALIZE
      CLL=0.35
      CLBL=6.
      CLBT=0.5
      CCLT=0.4
      CCLL=5.5
      HGT=0.06
```

```
      IDL=0
      IL=0
      IFIRST=0
      IT=0
C.....FIND INCREMENTS
    5 INCI=J2-J1
      INCJ=I1-I2
C.....FIND DISTANCE OF INTERSECTION ALONG SEGMENT
      D=(CONT-Z(I1,J1))/(Z(I2,J2)-Z(I1,J1))
      IF (D.GT.1.) D=1.
      IF (CONT-Z(I1,J1).EQ.0..OR.D.LT.0.) D=0.
C.....FIND NEW X,Y
      XN=XLEN*(J1-1+D*INCI)
      YN=YLEN*(NROWS-I1+D*INCJ)
      IF (IFIRST.EQ.0) GOTO 7
      DX=XN-X
      DY=YN-Y
      CLV=SQRT(DX**2+DY**2)
      IF (ABS(CLV).LT.0.00001) GOTO 3
      IF (DX.EQ.0.) DX=0.00001
      TANV=DY/DX
      CCLL=CCLL+CLV
C.....FIND TICKS IF NECESSARY
      IF (IL.NE.0) GOTO 8
      CCLT=CCLT+CLV
      NTICK=MIN0(INT(CCLT/CLBT),6)
      DO 9 ITICK=(CCLT-CLV)/CLBT+1,NTICK
      XTICK1(ITICK)=X+(CLBT*(ITICK)-CCLT+CLV)*DX/CLV
      YTICK1(ITICK)=Y+(CLBT*(ITICK)-CCLT+CLV)*DY/CLV
      XTICK2(ITICK)=XTICK1(ITICK)+DY*0.05/CLV
    9 YTICK2(ITICK)=YTICK1(ITICK)-DX*0.05/CLV
C.....DRAW LABEL IF NECESSARY
C.....CHECK FOR LABEL START AND DRAW CONTOUR UP TO LABEL
    8 IF (CCLL.LT.CLBL.OR.IDL.EQ.1) GOTO 2
      XL1=X+(CLBL-CCLL+CLV)*DX/CLV
      YL1=Y+(CLBL-CCLL+CLV)*DY/CLV
      CALL PLOT (IPLT,XL1,YL1,IPEN)
      IDL=1
C.....CHECK FOR LABEL END AND DRAW LABEL
    2 IF(IDL.NE.1) GOTO 7
      DSL=SQRT((XN-XL1)**2+(YN-YL1)**2)
      IF (DSL.LT.CLL) GOTO 4
      DX1=XL1-X
      IF (DX1.EQ.0.) DX1=0.00001
      ALPHA=ABS(ATAN((YL1-Y)/DX1)-ATAN(TANV))
      IF ((XL1-X)*DX.LT.0.) ALPHA=3.1416-ALPHA
      B=SQRT((XL1-X)**2+(YL1-Y)**2)
      P=B*COS(ALPHA)
      IF (P.GT.100.) P=100.
      RAD=P**2-B**2+CLL**2
      IF (RAD.LT.0.)RAD=0.
      A=P+SQRT(RAD)
      XL2=X+A*DX/CLV
      YL2=Y+A*DY/CLV
      ANG=ATAN((YL2-YL1)/(XL2-XL1))*180./3.1416
      IF (XL2.LT.XL1) GOTO 11
      XL3=XL1
      YL3=YL1
      GOTO 12
   11 XL3=XL2
      YL3=YL2
   12 XL3=XL3+(YL2-YL1)*SIGN(1.,XL2-XL1)/CLL*HGT/2.
      YL3=YL3-ABS(XL2-XL1)/CLL*HGT/2.
      CALL PLOT (IPLT,XL3,YL3,0)
      CALL CHARS (IPLT,ANG,HGT,0.,0.,1.,IPEN,5,LABEL)
      CALL PLOT (IPLT,XL2,YL2,0)
      IDL=0
      IL=1
      IT=0
      CCLL=CCLL-CLBL-CLL
      GOTO 8
C.....MOVE PEN (FIRST TIME) OR DRAW VECTOR
```

```
    7 IP=IFIRST*IPEN
      CALL PLOT (IPLT,XN,YN,IP)
    4 X=XN
      Y=YN
      IF (IDL.EQ.0) GOTO 3
      IT=IT+1
      XT(IT)=X
      YT(IT)=Y
C.....UNTAG SEGMENT AT END OF VECTOR AND RETURN IF NECESSARY
    3 IF (I1.EQ.I2.AND.(I1+INCI.GT.NROWS.OR.I1+INCI.LT.1)) GOTO 40
      IF (J1.EQ.J2.AND.(J1+INCJ.GT.NCOLS.OR.J1+INCJ.LT.1)) GOTO 40
      IF (I1.NE.I2.OR.I1.EQ.1.OR.I1.EQ.NROWS.OR.J1.GT.J2) GOTO 6
      IF (.NOT.tg(I1-1,J1)) GOTO 40
      tg(I1-1,J1)=.FALSE.
C.....INDICATE THAT FIRST MOVE IS DONE
    6 IFIRST=1
C.....FIND GEOMETRIC CONDITION OF NEXT VECTOR
      IA=0
      IB=0
      IF (Z(I2+INCI,J2+INCJ).LE.CONT) IA=1
      IF (Z(I1+INCI,J1+INCJ).GT.CONT) IB=2
      IC=IA+IB
      GOTO (10,20,30),IC
C.....NEITHER BACK NOR FRONT (I.E. SIDE ONLY)
      I1=I1+INCI
      I2=I2+INCI
      J1=J1+INCJ
      J2=J2+INCJ
      GOTO 5
C.....FRONT ONLY
   10 I1=I1+INCI-INCJ
      J1=J1+INCI+INCJ
      GOTO 5
C.....BACK ONLY
   20 I2=I2+INCI+INCJ
      J2=J2-INCI+INCJ
      GOTO 5
C.....BOTH BACK AND FRONT
   30 D1=(CONT-Z(I1+INCI,J1+INCJ))/(Z(I2+INCI,J2+INCJ)-Z(I1+INCI,
     + J1+INCJ))
      IF (D1-D) 10,10,20
C.....RETURN SEQUENCE
C.....DRAW STORED CONTOUR IF LABEL WASN'T FINISHED
   40 IF (IDL.EQ.0) GOTO 42
      DO 41 I=1,IT
   41 CALL PLOT (IPLT,XT(I),YT(I),IPEN)
C.....DRAW TICKS IF NO LABEL HAS BEEN DRAWN or label wasn't finished
   42 IF (IL.EQ.1.AND.IDL.EQ.0) RETURN
      DO 43 ITICK=1,NTICK
      CALL PLOT (IPLT,XTICK1(ITICK),YTICK1(ITICK),0)
   43 CALL PLOT (IPLT,XTICK2(ITICK),YTICK2(ITICK),IPEN)
      RETURN
      END
C-----------------------------------------------------------------------
      SUBROUTINE PLTCRS (IPLT,X,Y,IPEN)
C.....PLOTS CROSS AT POSITION (X,Y)
      CALL PLOT (IPLT,X-0.05,Y,0)
      CALL PLOT (IPLT,X+0.05,Y,IPEN)
      CALL PLOT (IPLT,X,Y-0.05,0)
      CALL PLOT (IPLT,X,Y+0.05,IPEN)
      RETURN
      END
C-----------------------------------------------------------------------
      SUBROUTINE PLTGRD (IPLT,IPEN)
C.....OPENS PLOT AND PLOTS GRID
      IMPLICIT COMPLEX (C)
      CHARACTER*72 IT
C.....FOR STANDARD FORTRAN CHANGE 'EXTENDED BLOCK' TO 'COMMON',
      COMMON IT,TO,TR,TD,DT,TB,TE,TID,TF,FLD,SWD,VIS,ROU,DV,CVC,CWV,
     & SDI(4),SDD(4),BD(4),FCO(5),
     & TI,LI,CPI(20),CVI(20),SLI(20,4),TEV,LEV,CPEV(20),
     & DX,N,M,NQ,QX(10,4),QY(10,4),QZ(10,4),
```

```
     & ZT(41,61),ZB(41,61),DZZ,L,
     & FEN(41,61),CVN(41,61),SUB(41,61),ZT1(41,61),NY
C.....OPEN PLOT
      XYLEN=6.87/MAX0(M-1,N-1)
      CALL OPENPLOT (IPLT,1,13.75,13.75,-1.,15.,-5.,11.)
      XMX=(M-1)*XYLEN
      YMX=(N-1)*XYLEN
C.....DRAW GRID OR BOX
      DO 1 J=1,M
      X=(J-1)*XYLEN
      CALL PLOT (IPLT,X,0.,0)
    1 CALL PLOT (IPLT,X,YMX,IPEN)
      DO 2 I=1,N
      Y=(I-1)*XYLEN
      CALL PLOT (IPLT,0.,Y,0)
    2 CALL PLOT (IPLT,XMX,Y,IPEN)
      CALL CLOSPLOT(IPLT)
      RETURN
      END
C----------------------------------------------------------------------------
      SUBROUTINE PLTNOD3 (IPLT,IPEN,K1,FENMIN,UAR,UDP)
C.....PLOTS VELOCITY VECTORS AT EACH NODE
C     K1=0 PLOTS VELOCITY ARROWS
C     K1=1 PLOTS FLOW ARROWS
      IMPLICIT COMPLEX (C)
      CHARACTER*72 IT
      CHARACTER*80 STRNG
C.....FOR STANDARD FORTRAN CHANGE 'EXTENDED BLOCK' TO 'COMMON',
      COMMON /BLK1/LBIT(50,41,61)
      COMMON IT,TO,TR,TD,DT,TB,TE,TID,TF,FLD,SWD,VIS,ROU,DV,CVC,CWV,
     & SDI(4),SDD(4),BD(4),FCO(5),
     & TI,LI,CPI(20),CVI(20),SLI(20,4),TEV,LEV,CPEV(20),
     & DX,N,M,NQ,QX(10,4),QY(10,4),QZ(10,4),
     & ZT(41,61),ZB(41,61),DZZ,L,
     & FEN(41,61),CVN(41,61),SUB(41,61),ZT1(41,61),NY
C.....SET PARAMETERS
      XYLEN=6.87/MAX0(M-1,N-1)
      SCALE=XYLEN/UAR
      SCALZ=0.
      IF (UDP.NE.0.) SCALZ=XYLEN/UDP
C.....OPEN PLOT AND DRAW GRID
      CALL OPENPLOT (IPLT,1,13.75,13.75,-1.,15.,-5.,11.)
C.....ENTER MAIN NESTED LOOPS (ONCE PER NODE)
      DO 1 I=1,N
      DO 1 J=1,M
      IF (FEN(I,J).LT.FENMIN) GOTO 1
      X=(J-1)*XYLEN
      Y=(N-I)*XYLEN
C.....PLOT CROSS IF VELOCITY IS 0.
      IF (CVN(I,J).EQ.(0.,0.)) CALL PLTCRS(IPLT,X,Y,IPEN)
C.....PLOT ARROW
      IF (K1.EQ.0) GOTO 4
      XD=REAL(CVN(I,J))*SCALE/2.*(FEN(I,J)*DZ)
      YD=AIMAG(CVN(I,J))*SCALE/2.*(FEN(I,J)*DZ)
      GOTO 5
    4 XD=REAL(CVN(I,J))*SCALE/2.
      YD=AIMAG(CVN(I,J))*SCALE/2.
    5 ZZ=DZ*SCALZ*FEN(I,J)
      CALL PLTARR (IPLT,X-XD,Y-YD,X+XD,Y+YD,ZZ,IPEN)
    1 CONTINUE
C.....PLOT POSITION OF SOURCES
      DO 2 K=1,LI
      X=REAL(CPI(K))/DX*XYLEN
      Y=AIMAG(CPI(K))/DX*XYLEN
      CALL PLTCRS (IPLT,X,Y,IPEN)
      WRITE (STRNG,10) K
   10 FORMAT (I2)
    2 CALL CHARS (IPLT,0.,0.1,0.,0.,1.,5,2,STRNG)
C.....PLOT SCALE
      ZZ=XYLEN
      CALL PLTARR (IPLT,(M-2)*XYLEN-.3,-0.15,(M-1)*XYLEN-.3,-0.15,ZZ,
     & IPEN)
```

```
      WRIYE (STRNG,13) UAR
   13 FORMAT (F4.0,' m/s')
      CALL PLOT (IPLT,(M-2)*XYLEN-.6,-0.30,0)
      CALL CHARS (IPLT,0.,0.1,0.,0.,1.,1,8,STRNG)
      CALL PLOT (IPLT,(M-2)*XYLEN-.7,-0.45,0)
      CALL CHARS (IPLT,0.,0.1,0.,0.,1.,1,11,'Veloc. Scale')
      CALL CLOSPLOT (IPLT)
      RETURN
      END
C----------------------------------------------------------------------
      SUBROUTINE PLTPAT3 (CPO,K,IPLT,IPEN)
C.....PLOTS PATHS OF INDIVIDUAL FLUID ELEMENTS
C     XYLEN=LENGTH IN INCHES OF CELL SIDE
      IMPLICIT COMPLEX (C)
      CHARACTER*72 IT
      COMMON IT,TO,TR,TD,DT,TB,TE,TID,TF,FLD,SWD,VIS,ROU,DV,CVC,CWV,
     & SDI(4),SDD(4),BD(4),FCO(5),
     & TI,LI,CPI(20),CVI(20),SLI(20,4),TEV,LEV,CPEV(20),
     & DX,N,M,NQ,QX(10,4),QY(10,4),QZ(10,4),
     & ZT(41,61),ZB(41,61),DZZ,L,CP(5000),CV(5000),SL(5000,4),
     & FEN(41,61),CVN(41,61),SUB(41,61),ZT1(41,61),SDER(41,61),
     & NY,G,DX2,DZ,DM,FF,FB,F3,F4,VST(4),FU(4),FD(4),TC(4),FAC
      COMMON /PPAT/F1,NYT,XYLEN,ICALL
      SAVE /PPAT/
C.....OPEN PLOT
      IF (K.NE.1.OR.NY.NE.0) GOTO 2
      F1=3.1E+07
      IF (ICALL.NE.1) IPLT=2
      ICALL=1
      NYT=TE*F1/DT/TB
      XYLEN=6.87/MAX0(M-1,N-1)/DX
      CALL OPENPLOT (IPLT,1,13.75,13.75,-1.,15.,-5.,11.)
    2 X1=REAL(CPO)*XYLEN
      IF (X1.LT.0.) GOTO 1
      Y1=AIMAG(CPO)*XYLEN
      X2=REAL(CP(K))*XYLEN
      IF (X2.LT.0.) GOTO 1
      Y2=AIMAG(CP(K))*XYLEN
      IF (CPO.EQ.CP(K)) CALL PLTCRS (IPLT,X1,Y1,IPEN)
      IF (CPO.NE.CP(K)) CALL PLTARR (IPLT,X1,Y1,X2,Y2,0.,IPEN)
    1 IF (K.LT.L.OR.NY.LT.NYT) RETURN
      CALL CLOSPLOT (IPLT)
      IPLT=IPLT+1
      RETURN
      END
C----------------------------------------------------------------------
      SUBROUTINE READGU3 (IF1,IF2)
C.....READS FROM UNFORMATTED GRAPHIC FILE
C     IF1=1 IF END OF FILE IS FOUND
C     IF2=1 DON'T READ GENERAL DATA
      IMPLICIT COMPLEX (C)
      CHARACTER*8 BLANK
      DIMENSION RD(61)
C.....FOR STANDARD FORTRAN CHANGE 'EXTENDED BLOCK' TO 'COMMON',
      CHARACTER*72 IT
C.....FOR STANDARD FORTRAN CHANGE 'EXTENDED BLOCK' TO 'COMMON',
      COMMON /BLK1/LBIT(50,41,61)
      COMMON IT,TO,TR,TD,DT,TB,TE,TID,TF,FLD,SWD,VIS,ROU,DV,CVC,CWV,
     & SDI(4),SDD(4),BD(4),FCO(5),
     & TI,LI,CPI(20),CVI(20),SLI(20,4),TEV,LEV,CPEV(20),
     & DX,N,M,NQ,QX(10,4),QY(10,4),QZ(10,4),
     & ZT(41,61),ZB(41,61),DZZ,L,
     & FEN(41,61),CVN(41,61),SUB(41,61),ZT1(41,61),NY
      DATA BLANK/'        '/
      IF1=0
      IF (IF2.EQ.1) GOTO 19
C.....READ DATA FILE
      READ (22,ERR=10,END=20)
     & IT,TO,TR,TD,DT,TB,TE,TID,TF,FLD,SWD,VIS,ROU,DV,CVC,CWV,
     & (SDI(K),K=1,4),(SDD(K),K=1,4),(BD(K),K=1,4),
     & (FCO(K),K=1,5),TI,LI
      IF (LI.GT.0) READ (22,ERR=10,END=20)
```

```
     & (CPI(K),CVI(K),(SLI(K,L1),L1=1,4),K=1,LI)
      READ (22,ERR=10,END=20) TEV,LEV
      IF (LEV.NE.0) READ (22,ERR=10,END=20) (CPEV(LEVI),LEVI=1,LEV)
      READ (22,ERR=10,END=20) DX,N,M,NQ
      IF (NQ.GT.0) READ (22,ERR=10,END=20)
     & ((QX(K,L1),QY(K,L1),QZ(K,L1),L1=1,4),K=1,NQ)
      RETURN
C.....READ DATA FOR CURRENT CYCLE
   19 READ (22,ERR=10,END=99) TO
      DO 28 I=1,N
      IF (TO.EQ.0.) GOTO 27
      DO 1 J=1,M
    1 ZT1(I,J)=ZT(I,J)
   27 READ (22,ERR=10,END=20) (ZT(I,J),J=1,M)
   28 CONTINUE
      DO 2 I=1,N
      READ (22,ERR=10,END=20) (ZB(I,J),J=1,M)
      IF (TO.NE.0.) GOTO 2
      DO 35 J=1,M
   35 ZT1(I,J)=ZB(I,J)
    2 CONTINUE
C.....READ LITHOLOGY
      READ (22,ERR=10,END=20) DZZ
      DO 29 I=1,N
      DO 29 J=1,M
   29 READ (22,ERR=10,END=20) KL,(LBIT (K,I,J),K=1,KL)
      DO 9 I=1,N
    9 READ (22,ERR=10,END=20) (FEN(I,J),J=1,M)
      READ (22,ERR=10,END=20) EA
      DO 31 I=1,N
      READ (22,ERR=10,END=20) (RD(J),J=1,M)
      DO 31 J=1,M
   31 CVN(I,J)=CMPLX(RD(J),0.)
      DO 12 I=1,N
      READ (22,ERR=10,END=20) (RD(J),J=1,M)
      DO 12 J=1,M
   12 CVN(I,J)=CVN(I,J)+CMPLX(0.,RD(J))
      RETURN
   10 WRITE (6,11)
   11 FORMAT (' SUBROUTINE READGU FOUND ERROR IN GRAPHIC FILE')
      STOP
   20 WRITE (6,21)
   21 FORMAT (' SUBROUTINE READGU FOUND EOF IN GRAPHIC FILE')
      STOP
   99 IF1=1
      RETURN
      END
C---------------------------------------------------------------------
      SUBROUTINE SEDNO13 (I,J,K,LIT)
C     I=ROW OF NODE
C     J=COL OF NODE
C     K=CELL, IF 0, NEGATIVE OR > 800, ERROR IS RETURNED
C     LIT=LITHOLOGY: 1, 2, 3, OR 4. IF 0 or NEGATIVE, lit is returned
      IMPLICIT COMPLEX (C)
      CHARACTER*72 IT
      COMMON /BLK1/LBIT(50,41,61)
      COMMON IT,TO,TR,TD,DT,TB,TE,TID,TF,FLD,SWD,VIS,ROU,DV,CVC,CWV,
     & SDI(4),SDD(4),BD(4),FCO(5),
     & TI,LI,CPI(20),CVI(20),SLI(20,4),TEV,LEV,CPEV(20),
     & DX,N,M,NQ,QX(10,4),QY(10,4),QZ(10,4),
     & ZT(41,61),ZB(41,61),DZZ,L,
     & FEN(41,61),CVN(41,61),SUB(41,61),ZT1(41,61),NY
C.....ERODE
      IF (K.LE.800.AND.K.Gt.0.and.lit.ge.0.and.lit.le.4) GOTO 2
      OPEN (24,FILE='for24.d',FORM='FORMATTED')
      rewind(24)
      WRITE (24,11) TO,TR,L
   11 FORMAT (1X,' TIME=',F15.7,' YEARS OF ',F15.7,' YEARS'/
     & ' L=',I6)
      WRITE (24,12) K,lit
   12 FORMAT (' SED. COLUMN < 1 OR > 800, K=',I4,' LIT=',I1)
      CLOSE (24)
```

```
      STOP
    2 LW=(K+15)/16
      LB=2*K+14-16*LW
      IF (LIT.GT.0) GOTO 1
      LIT=IBITS(LBIT(LW,I,J),LB,2)+1
      RETURN
C.....DEPOSIT
    1 IF((LIT-1)/2.EQ.1) THEN
      LBIT(LW,I,J)=IBSET(LBIT(LW,I,J),LB+1)
      ELSE
      LBIT(LW,I,J)=IBCLR(LBIT(LW,I,H),LB+1)
      ENDIF
      IF(MOD(LIT,2).EQ.0) THEN
      LBIT(LW,I,J)=IBSET(LBIT(LW,I,H),LB)
      ELSE
      LBIT(LW,I,J)=IBCLR(LBIT(LW,I,J),LB)
      ENDIF
      RETURN
      END
C-----------------------------------------------------------------------
```

Glossary

This glossary defines most of the important technical terms used in this volume. Some of the definitions are specific for their context in this volume, but others are in accord with their general scientific use.

Anisotropic: A medium in which physical properties vary according to the direction in which they are measured.

Armor layer: A layer of relatively coarse material formed naturally on a stream bed by the removal of finer material.

Bed load: Sediment that moves by jumping, rolling, or sliding, and remains within two or three grain diameters above the bed (see also *suspended load*).

Bed-material load: Part of the sediment load composed of particles of the same size as those in the bed (see also *wash load*).

Cell: The portion of a surface or of a volume representing an elementary constituent of a grid. Cells in SEDSIM are either squares or rectangular blocks.

Cell boundary: The line or surface defining the border or edge of a cell.

Continuity equation: An equation that states that mass cannot be lost or gained overall but only transferred from one location to another.

Deterministic: A process whose outcome or behavior can be predicted from previous or initial conditions.

Eulerian: A method of defining flow in which properties of the flow are given at points that are fixed in space.

Finite-difference procedures: Numerical techniques that employ regular grids and approximate differential equations with algebraic expressions that express differences between values of variables at grid nodes.

Finite-element procedures: Numerical techniques that employ grids that may be irregular and approximate differential equations with algebraic expressions that stem from integrating a variable throughout a cell.

Fixed surface: A surface that does not change shape or position.

Flow: The type of motion characteristic of fluids, usually represented by a vector field that indicates velocity at each point. Also, a moving volume of fluid.

Flow-velocity profile: A function representing the distribution of velocities along a vertical line through the flow.

Fluid element: A movable point representing a volume of fluid.

Free surface: A surface that bounds a fluid and is free to move as determined by forces within the fluid. Usually applied to the air-water interface in an open channel.

Grid: A mesh by which a surface or a volume can be partitioned into cells.

Grid element: A cell or a node.

Grid, fixed: A grid that does not change shape with time.

Grid, regular: A grid in which all cells have the same shape and size.

Homogeneous: A medium in which physical properties are the same for any portion of the medium.

Homogeneous sediment: An aggregate of particles that has the same physical properties everywhere when observed in bulk, but not necessarily consisting of particles that are all identical to each other.

Isotropic: A medium in which physical properties are the same regardless of direction.

Iteration: Each step within a number of repetitive steps used in obtaining a numerical solution to a set of differential equations.

Lagrangian: A method of defining flow in which properties of the flow are given at points that move with the fluid.

Model: A representation of a real system in the form of either a scaled physical reproduction or a conceptual or theoretical description, often embodied as a computer program.

Modeling: The process of building a model to reproduce the behavior of a real system.

Momentum equation: An equation balancing specific factors that affect the changes in the motion of an elementary mass of fluid.

Node or grid point: The point occupied by one or more grid cell vertices.

Random: A process whose outcome or behavior cannot be completely predicted from previous or initial conditions and which therefore depends on unpredictable factors.

Sediment transport equation: An equation defining how much sediment is carried by the flow according to upstream supply, flow conditions, and bed composition.

Rigid surface: A surface that bounds a fluid and does not move appreciably as compared to the flow velocity. A channel bed, for example, is a rigid surface because, although erosion or deposition can cause it to move, it moves upward or downward extremely slowly (see also *fixed surface*).

Sediment type: A specific fraction of a mixture of sediment composed of particles that are alike in physical characteristics, including size, shape, and density.

Simulation: The process of conducting experiments with a model, usually with the purpose of understanding or predicting the behavior of the system that the model reproduces.

Suspended load: Sediment that is supported by upward components of turbulent motion and remains in suspension for an appreciable time. Bed load and suspended load constitute total load.

System: A mechanism that is characterized by a set of components and the relationships and interactions between the components.

Total load: The total amount of sediment that is being moved by the flow. Total load is equal to bed load plus suspended load, or bed-material load plus wash load.

Uniform sediment: Sediment that is composed of a single type and represented by particles that are alike in size, shape, and density.

Variable, continuous: A variable that takes values at every point on a line, on a surface, or in space.

Variable, discrete: A variable that takes values at a set of points that are separated from each other.

Viscosity: A property of a fluid defining the fluid's resistance to shear.

Wash load: The part of the sediment load composed of particles that are smaller than the particles composing the bed. The amount of wash load is usually determined by upstream conditions. The wash load plus the bed-material load is equal to the total load.

References

Bates, C. C., 1953, Rational theory of delta formation, *Am. Assoc. Petroleum Geol. Bull.* **37**: 2119–2162.

Bonham-Carter, G., and A. J. Sutherland, 1968, *Mathematical Model and FORTRAN IV Program for Computer Simulation of Deltaic Sedimentation*, Computer Contribution No. 24, Kansas Geological Survey, Lawrence, Kansas, 56p.

Buneman, O., et al., 1980, Principles and capabilities of 3-D, E-M particle simulations, *Jour. Computational Physics* **38**: 1–44.

Carter, R. D., C. G. Mull, K. J. Bird, and R. B. Powers, 1977, The Petroleum Geology and Hydrocarbon Potential of the Naval Petroleum Reserve No. 4, North Slope, Alaska, U.S. Geol. Survey Open-File Report 77-475, 61p.

Cheng, R. T., 1983, Euler-Lagrangian computations in estuarine hydrodinamics, in C. Taylor, J. A. Johnson, and R. Smith, eds., *Proceedings of the Third International Conference on Numerical Methods in Laminar and Turbulent Flow*, Pineridge Press, pp. 341–352.

Colby, B. R., 1964, Practical computations of bed-material discharge, *Jour. Hydraulics Division, Am. Soc. Civil Engrs.* **90**, n. HY2.

Du Boys, M. P., 1879, Etudes du regime et l'action exercee par les eaux sur un lit a fond de graviers indefiniment affouliable, *Annales des Ponts et Chausses, Ser . 5* **18**: 141–195.

Harbaugh, J. W., and G. Bonham-Carter, 1970, *Computer Simulation in Geology*, Wiley-Interscience, New York, 575p.

Harlow, F. H., 1964, The particle-in-cell computer method for fluid dynamics, in B. Alder, ed., *Computational Physics*, vol. 3, Academic Press, New York, pp. 319–343.

Hockney, R. W., and J. W. Eastwood, 1981, *Computer Simulation Using Particles*, McGraw-Hill, New York, 523p.

Kalinske, A. A., 1947, Movement of sediment as bed load in rivers, *Trans. Am. Geophys. Union* **28**: 615.

Lane, E. W., and E. J. Carlson, 1953, Some factors affecting stability of canals constructed in coarse granular materials, *Proc. Minnesota Internat. Hydraulics convention, Joint Meeting of the International Association for Hydraulics Research and Hydraulics Division*, Am. Soc. of Civil Engrs. (August 1953), p. 37.

Laursen, E. M., 1956, The application of sediment transport mechanics to stable channel design, *Jour. Hydraulics Division, Am. Soc. Civil Engrs.* **82**: n. HY4.

Laursen, E. M., 1958, The total sediment load of streams, *Proc. Am. Soc. Civil Engrs.* **84**: 1530–1531.

Manning, R., 1890, Flow of water in open channels and pipes, *Trans. Inst. Civil Engrs. (Ireland)* **20**.

Meyer-Peter, E., and R. Muller, 1948, Formulas for bedload transport, *Proceedings, Third Meeting of Intern. Assoc. Hydr. Res., Stockholm*, pp. 39–64.

Mitchum, R. M., Jr., 1977, Seismic stratigraphy and global changes in sea level, Part 11: Glossary of terms used in seismic stratigraphy, in *Seismic Stratigraphy—Applications to Hydrocarbon Exploration*, Am. Assoc. Petroleum Geol. Mem. **26**: 205–212.

Perlmutter, M. A., 1987, personal communication, letter dated April 10, 1987.

Prandtl, L., 1930, *Ergebnisse der Aerodynamischen Versuchsanstalt zu Goettingen*, R. Oldenburg, Munich and Berlin, p. 32.

Sangree, J. B., and J. M. Widmier, 1977, Seismic stratigraphy and global changes in sea level, Part 9: Seismic interpretation of clastic depositional facies, in C. E. Payton, ed., *Seismic Stratigraphy—Applications to Hydrocarbon Exploration*, Am. Assoc. Petroleum Geol. Mem. **26**, Tulsa, Oklahoma.

Saxena, R. S., 1976, *Modern Mississippi Delta—Depositional Environments and Processes: A Guidebook Prepared for the Am. Assoc. Petroleum Geol./Soc. Econ. Paleontologists and Mineralogists Field Trip—Mississippi Delta Flight*, Am. Assoc. Petroleum Geol., 125p.

Scott, N., 1986, Modern vs. Ancient Braided Stream Deposits: A Comparison between Simulated Sedimentary Deposits and the Ivishak Formation of the Prudhoe Bay Field, Alaska, M.S. Thesis, Department of Applied Earth Sciences, Stanford University.

Shields, I. A., 1936, Anwendung der Ähnlichkeitsmechanik und der Turbulenzforschung auf Geschiebebewegung, *Mitteilungen der Preussischen Versuchsanstalt für Wasserbau und Schiffbau*, Heft 26, Berlin. (Available also as translation by W. P. Ott and J. C. van Uchelen, S.C.S. Cooperative Laboratory, California Institute of Technology, Pasadena, Calif.)

Tetzlaff, D. M. and J. W. Harbaugh, 1985, Computer mapping of seismic reflectors in the coastal region of the National Petroleum Reserve in Alaska, *Math. Geol.* **17**: 445–480.

Trowbridge, A. C., 1930, Building of Mississippi delta, *Am. Assoc. Petroleum Geol. Bull.* **14**: 867–901.

Vanoni, V. A., and G. N. Nomicos, 1960, Resistance properties of sediment-laden streams, *Trans. Am. Soc. Civil Eng.* **125**: 1140.

Index

MIX
Papier aus verantwortungsvollen Quellen
Paper from responsible sources
FSC® C105338

If you have any concerns about our products,
you can contact us on
ProductSafety@springernature.com

In case Publisher is established outside the EU,
the EU authorized representative is:
Springer Nature Customer Service Center GmbH
Europaplatz 3, 69115 Heidelberg, Germany

Printed by Libri Plureos GmbH
in Hamburg, Germany